The Web Beneath
the Waves

COLUMBIA GLOBAL REPORTS
NEW YORK

The Web Beneath the Waves

The Fragile Cables That Connect Our World

Samanth Subramanian

Undersea Cables, 2000

Undersea Cables, 2025

This book is supported by the Alfred P. Sloan Foundation under Grant No. G-2025-25211.

Published by Columbia Global Reports
91 Claremont Avenue, Suite 515
New York, NY 10027
globalreports.columbia.edu

Library of Congress Cataloging-in-Publication Data Available Upon Request

ISBN 979-8-987053-78-2 (paperback)

Cover by Kelly Winton
Map design by Jeffrey L. Ward
Author photograph by Chinky Shukla

Printed in the United States of America

CONTENTS

For Vedant and Shravan,
who grew up
never knowing a world without the internet

The Web Beneath
the Waves

A Cable Snaps

For a while, Sam Vea had been smelling sulfur on the air—only mildly infernal, like a distant sniff of hell, but sulfur nonetheless. Still, on that Saturday evening, when the explosion happened, he sat up in fright. It sounded so near he thought some cataclysm had occurred right there, in his neighborhood. The windows trembled. The curtains fell off. Vea peeked out of his house but saw nothing destroyed or on fire, so he looked at his wife and said: "This has to be the volcano."

Vea and his wife live in Tofoa, which, if you squint and picture Tonga's main island of Tongatapu as a long, medieval shoe, lies just below the instep, on a gentle rise of earth. They'd just returned home after dropping their daughters at a birthday party, but now Vea dashed into his van to go collect them. On the way back, the road filled with cars scurrying away from the sea, and tiny pebbles fell from the sky. Not that long ago, curious to see what a big volcanic eruption looked like, Vea had watched *Dante's Peak* on Netflix. In the movie, he remembered now, a white-hot rock had punched through the roof of a truck and killed Pierce

Brosnan's partner, so he pulled over to wait out the traffic. The skies grew mottled with dust and ash. Drivers got out, took off their shirts, and wiped their windshields down so that they could see the road ahead. When they reached home, after two and a half hours, Vea sent his children to hide under the bed.

The volcano, with the grand, rolling name of Hunga Tonga-Hunga Ha'apai, lies forty miles north of Tongatapu—mostly under the Pacific Ocean but with two spits of land showing above the water, like the ears of a drowned cat. A volcanic crater, so low that it skimmed the water, linked these outcrops. In Tongan lore, these islets were among the *makafonua,* or land-stones, of Hikuleo, the divine maker of all the earthly terrain that is uneven and that jumps backward and forward—which is to say, wracked by earthquakes. Since its several brief eruptions the previous month, December 2021, Hunga Tonga-Hunga Ha'apai had continued to gurgle and churn. That Saturday, in mid-January, 2.4 cubic miles of sediment and molten rock shot through its mouth with the force of what scientists call a "magma hammer," sending a plume of ash at least thirty-five miles up into the atmosphere and possibly higher, all the way into space. It was the largest atmospheric explosion that modern instruments had recorded, outdoing any nuclear bomb ever detonated. They heard the sound in Alaska. Seven and a half thousand miles away, in the south Indian city of Chennai, meteorologists measured an abrupt spike in atmospheric pressure. It was Hunga Tonga-Hunga Ha'apai, doing its thing.

On his drive home, Vea had called relatives in the US through Facebook Messenger, to let them know he was all right. At some point during their conversation, the line cut out. He assumed the network was overloaded by everybody getting online at the

16   same time. "This is usually a problem for us," he told me. Vea, DHL's agent in Tonga, is the president of the Tonga Chamber of Commerce & Industry, and we met in his spare, sunny office in the capital of Nuku'alofa, three streets from the Pacific. The curtains were red, and the sun filtered through in a dull watermelon light. Vea wears a perpetual expression of mirth, and it was difficult to imagine him as worried as he was on the day the volcano blew up, sitting in his van in the middle of a rain of ash, staring at his suddenly useless phone. He decided he'd try his relatives later, after the traffic online subsided. At home, though, the power was out, and he couldn't charge his phone, so it was only the next morning, when he tuned in to Radio Tonga, that he learned that the country had lost its internet altogether—and with it, all its means of reaching the world beyond the wide, silent water.

A year and a half after the eruption, no one still knew precisely what happened on the ocean floor that Saturday. But the marine geologist Mike Clare, at the National Oceanography Centre in Southampton, had, along with his team, pored over enough sonar readings and sediment samples to build a theory.

    In the abyssal depths of the ocean, a data cable is a scrawny, lightly clothed thing. Its core consists of fibers of glass, each no thicker than a human hair, through which light transmits information at roughly 125,000 miles per second. Around the fibers, there is first a casing of steel for protection, then another of copper to carry electricity to keep the light moving, and then a final sheath of nylon soaked in tar. All this swaddling may sound like plenty of protection, but the layers are all thin, and the final product is—to use the image I heard most often from people in the subsea cable industry—no fatter than a garden hose. Some veterans draw this comparison in tones of both guilt and marvel, as if they're ashamed of consigning this delicate thing to the sea and astonished that it performs (for the most part) as reliably and unfussily as it does.

Beneath miles and miles of water, these cables sit on the sea floor, conducting 95 percent of all the world's internet traffic. Humans have laid 870,000 miles of fiber optic cables under the ocean, connecting and reconnecting the eyelets on our shorelines, lacing the earth tightly together. Cables set out from places like Crescent Beach in Rhode Island, Wall Township in New Jersey, and Island Park in New York, and end in places like Penmarch in France, Bilbao in Spain, and Bude in the United Kingdom. Cables run between Milton in Newfoundland and Nuuk in Greenland, and between Toamasina in Madagascar and Saint-Marie in Réunion, and between Tarakan and Manado in Indonesia. Among the shortest cables anywhere is the forty-one-mile tyke between the Isle of Man and Northern Ireland; among the longest is 2Africa, which spans 28,000 miles and has branches into too many countries to list here. One brief cable stretches between Guningtou in Taiwan and Dadeng Island in China, countries that otherwise want little to do with each other. Another cable starts in the state of Washington and lands in Alaska, because sometimes it's easier or cheaper to put down cable under the sea than over land. Longyearbyen, in Svalbard, has not one but two cables linking it to the Norwegian mainland. Even tiny St. Helena, population 4,439, got hooked up to an undersea cable in 2023. Cables have acronymic names like SAFE and AAE, or functional names like Greenland Connect and Pacific Crossing-1, or mythopoeic names like JUPITER and Honomoana, or historical names like Leif Erikson and Grace

18   Hopper. There are roughly 550 such submarine cables around the planet, and more are being built every day. A Finnish company planned to spend a billion or so dollars to lay cable under the Arctic Ocean—a task made easier by how rapidly its ice cover is melting. Upon completion, the cable was designed to shave twenty to sixty milliseconds off the speed of trades made by banks in Tokyo and London. For now, Antarctica is the only major uncabled landmass on Earth, but it won't be for long. The US has plans.

The cable that connects Tongatapu to Fiji and thence to the world, 515 miles long and part of a cable network called Southern Cross, was switched on in 2013. A 250-mile domestic cable between Tongatapu and the northern island of Vava'u began operating in 2018. In general, this pocket of the Pacific is a rough neighborhood for undersea cables. "Ideally, you'd route around seamounts—but what happens if your island itself is a volcano, as happens in the Caribbean or the Pacific?" Clare said. "And if you're looking at standoff distances"—by which he meant the gap that a cable maintains between itself and a hazard—"if you make your standoff distance one hundred kilometers in an area like this, you've already hit the next volcano." The cable could have been plotted to land not in the west of Tongatapu but in the east—but that would have meant negotiating a steep underwater slope with plenty of canyons and a reputation for earthquakes. "So that's a really gnarly place to lay a cable. You have to go with the least worst option."

Clare's theory about the post-eruption outage ran like this. When Hunga Tonga-Hunga Ha'apai shot its innards into the atmosphere, the dense rock and sediment fell back into the ocean at a furious pace, hit the flanks of the volcano, and

rocketed down its slopes. "It's like an avalanche, and the material shot along like a log flume in a theme park," Clare said.
Along the way, the flow of volcanic debris gathered more
mass still, so that when it met the domestic cable just a few
miles away, it was moving as fast as a speeding car on a highway. The contest was over before it began. The rock and mud
slammed into a sixty-five-mile section out of the domestic
cable and buried it under tens of feet of sediment. A different part of the flow, or perhaps the same one, sliced fifty-five
miles of cable out of the middle of the international link to
Fiji. In a cold-storage room, Clare showed me a vertical sample of the seabed that a ship had retrieved after the eruption. It
was encased in a long rectangular frame, and reading from bottom to top, it held first the mud of the original seabed the color
of milk chocolate; then a few inches of black, coarse volcanic
material; and finally the softer silt and sediment that fell back
to earth to be deposited on the ocean floor. The sample had
been taken seventy miles away from the volcano, but even at
that distance, the third layer was several feet thick. It felt cool
and moist to my touch, like new clay.

By the time Clare woke up in Southampton on the day of the
eruption, his Twitter feed was already awash in discussions and
satellite imagery. It took him, and most of the outside world, the
better part of a day to realize that Tonga had lost its internet. In
theory, he could have watched that happen in real time. In one of
the strange involutions of the modern age, we go onto the internet to see what's the matter with the internet, and there's a website, Clare said, where you can see the cardiogrammed tale of the
domestic and then the international cable giving up the ghost.
"Basically, the eruption happens, and fifteen minutes later,

20   there's a drop to about half of what the original internet traffic was, and then about an hour after the eruption, it flatlines."

That moment was when Sam Vea's cell phone gave out. Landlines died as well, because in Tonga, as in many other countries, even ordinary telephone calls are now routed through data cables. Another modern oddity: In Southampton, Clare could look at satellite photos and see that the eruption had blessedly left Vava'u, Tongatapu, and other islands in the Tongan archipelago intact, but Tongans themselves couldn't be sure of that. They had no way of communicating with each other, no way of learning about the condition of other parts of their own, small country. "For a week, I didn't know what happened to my family on Tongatapu," one man in Vava'u told me. "I have a brother in Nuku'alofa. I had to assume he was okay." Another said: "We thought Tongatapu was obliterated. There was just no way to know differently." In this age, a country offline may as well not exist at all.

We inhabit the internet in an odd, paradoxical state. It is everywhere, available to us all the time, whenever we desire it, like the air we breathe. But the ability to use the internet wirelessly permits us to forget not only its materiality—bottomless quantities of metals and plastics poured and cast into wires, routers, data centers, servers, towers, and repeaters—but also its centrality in our lives. We're lulled into believing that the internet is only a vehicle for emails, selfies, Zoom meetings, and websites that linger too long on unread browser tabs. That the very apparatus of twenty-first-century life relies on the internet is rendered visible to us only when something snaps, like the sole cable running to Tonga.

The first fallout was communication, of course. In the aftermath of a disaster, even the humble text message assumed grave importance. *Are you safe? Is your house still standing? Is the water safe to drink?* Tonga runs on Facebook Messenger, particularly on its outer islands, where the phone service is spotty, and without it, people had to take to the road—or the sea, or the air—to find anything out. The British High Commission's staff drove around to the homes of British citizens to check on them. From some of the outer islands, village and town officers came to Tongatapu by boat to deliver news of their communities. Tonga Communications Corporation, a government-owned body, had grown so used to the permanence of the undersea cable that it had dismantled its satellite equipment and let its subscription lapse. To renew it, they had to go online and make a payment—which, of course, they couldn't do. In any case, the sky was so thick with ash for a few days that satellite signals would have failed to penetrate it. Australia and New Zealand had to send reconnaissance planes over the islands so that their pilots could eyeball the extent of the damage.

Commerce broke down. Since this was the middle of the COVID-19 pandemic, DHL was flying only one plane a week to Tonga, but without the internet, Vea couldn't file or receive manifests online. ATMs went dead, because banks couldn't check how much money their customers had in their accounts—and that, in an economy still accustomed to cash, immediately put livelihoods in danger. Owners of fisheries and farms of squash and breadfruit were unable to fill out the compliance and quarantine forms needed to export their produce. Tongans living overseas couldn't wire funds home to help their families—and at the time, foreign remittances made up 44 percent of the

22   country's GDP. Optimistic tourists, anticipating the resump-
tion of post-COVID life in a few months, couldn't book hotels
or bed-and-breakfasts in Tonga. In countries in the northern
hemisphere, where lockdowns had been suspended, Vea told me,
the Tongan diaspora was still celebrating formal occasions like
weddings—and funerals, thanks to COVID. Ordinarily, they'd
search Facebook for vendors in Tonga who made and sold the
appurtenances of such ceremonies: the *ngatu*, a printed cloth
made out of the bark of a mulberry tree, or the *ta'ovala*, a mat
woven out of pandanus leaves and worn around the waist like
a girdle. They'd pay online, and Vea or another freight agent
would ship the materials abroad. Every link in that chain ceased
to function. The weddings and funerals went ahead, but incom-
plete. They couldn't even be livestreamed to relatives back home.
The cable break interrupted not just life but death as well.

When a wave of COVID hit Tonga, children were told to
stay home, but they couldn't attend classes online. Instead,
Vea's kids listened to lessons sporadically broadcast on the
radio. On Vava'u, people tried not to fall too grievously ill. Brian
Meikle, an American who has lived on the island for more than
a decade, described the local hospital as the kind that you went
to only for "super-minor things. If you break a bone, you go to
Tongatapu." Without access to the internet, though, the hospital
in Vava'u couldn't send its counterpart on Tongatapu any kind
of references—no doctors' notes or medical histories or requests
for transfers. Karen Stone, who runs an environmental nonprofit
on Vava'u, recalled two friends who died in a tragic accident in
the internet-less period after the eruption. "They were working
on their boat, and there must have been a live wire somewhere,"
she said. When they touched a metal part of their boat's frame,

they were electrocuted. "But we had no phones and no internet, and people couldn't check in on each other. They were found only ten days later." Some of the ramifications of the lack of internet were unexpected. Stone's house, wired to run on solar energy, suddenly lost power one day. It turned out her battery, which constantly trawled the internet for software updates, had bricked up when it couldn't get online.

When I first learned about Tonga's internet outage, I thought its citizens must have been hurled back to the 1990s. But, in fact, the internet has replaced so many other technologies—and Tonga was receiving so few visitors thanks to the pandemic— that the country was catapulted further back still, to a time before the telegraph and scheduled flights reached these parts of the Pacific. With the fracture of a single cable, Tonga was plunged into the kind of isolation it hadn't seen in more than a century.

Tonga's cable was cut by a freakish act of nature—but a volcanic eruption is only one of the myriad, complex perils facing the planet's network of underwater data cables. Some of these are also marine or geological: landslips, strong currents, the very infrequent shark attack. Others are the product of human accidents, such as a poorly dropped anchor or a fishing boat trawling too close to a cable. These categories of dangers have attended cables since the mid-nineteenth century, when humans first decided to lay a telegraph wire at the bottom of the sea.

The latest risks, minted over the last decade, are corporate misbehavior and geopolitical strife. Increasingly, subsea cables are commissioned and owned by a very small group of private tech firms like Google and Meta—American companies that can easily foot a new cable's bill that runs into the hundreds of

24   millions of dollars. These are also, not coincidentally, companies that make their money off the data of internet users, which gives them an acute vested interest in the traffic flowing through the cables they lay. In parallel, the great powers have realized that data cables in international waters are ripe targets—either for unsubtle duress or outright sabotage—precisely because they're both so important and so remote. The US and China routinely attempt to thwart each other's cable projects: permissions denied, contracts stymied, the intrigue as coiled and convoluted as tangles of cable. So extreme is this silent conflict that China has begun building the rudiments of its own, parallel network of subsea infrastructure—and is simultaneously experimenting with ultra-secure communications satellites designed to be immune to hacking. At a demonstration in March 2025, China set up a quantum-key encrypted link between Beijing and South Africa, with the promise to launch a global network by 2027.

    Incidents of deliberate damage have become so common that they've begun to feature in thriller novels—as in Harriet Crawley's The Translator, in which Russia plots to slice through the cables running to southern England. In the spring of 2024, three cables passing through the Red Sea were cut, allegedly by the maritime activities of Houthi militants in Yemen. The following year, China unveiled a tool that can be carried on submersibles to slice through armored cables at depths of up to thirteen thousand feet. Europe's nations have grown more and more convinced that their subsea cables are being cut by ships belonging to Russian or Chinese "shadow fleets"—civilian vessels doing their governments' bidding. In January 2025, Swedish investigators boarded one such ship, suspecting it of damaging a data cable between Latvia and Sweden—the fourth instance, within a span of three months, of a cable cut in the Baltic Sea. Around the same time, NATO launched a mission named "Baltic Sentry," recruiting warships, drones, and patrol airplanes to the task of safeguarding its cables. The UK now has warships prowling its waters to protect its cables. "There is reason for grave concern," Mark Rutte, NATO's secretary general, said.

What happened to Tonga could, in theory, happen to anyone—even to the world's biggest, wealthiest nations. The two US coasts, for instance, may be far more thickly connected with cables than Tongatapu, but all those cables do eventually run into the inky depths of the deep ocean, where they're protected by neither military nor legal muscle. The world today has come to depend utterly on these cables—and in tandem, these cables have grown more and more exposed to the whims of rogue corporate and national actors. The future of the internet, in part, will entail the weaponization of its submarine cable systems. Information, after all, is wealth and power—not only in how you use it, but how you can throttle it.

Tonga's exile from the internet was disruptive enough, but in a place closer to the heart of the global economy, the consequences of a blackout would be more dire still. As of this writing, fifteen submarine cables tie Taiwan to the rest of the world. Picture some tectonic or even man-made event that cuts them all. (This isn't as wild as it sounds. In 2006, an earthquake ruptured eleven cables traversing an undersea canyon off Taiwan's southwestern coast.) Submarine cables carry $10 trillion in financial transactions daily; without them, Taiwan's stock exchange and banks would seize up, and their chaos would spill over its borders into other economies. Taiwan's semiconductor foundries, keeping the world flush with microchips, rely on the cloud just as

26 companies elsewhere do; were they to go offline, even for a matter of days, the global electronics industry would stall. Taiwan's military, forever vigilant about Chinese incursions into its air and waters, would struggle to gather intelligence or even to communicate with itself. In the absence of the internet, Taiwan's government wouldn't be able to verify or counter foreign misinformation. Taiwan's airport, through which 48 million passengers pass every year, would still have its radios, but its air traffic controllers often speak to airplanes and to each other via channels on the internet. Every one of Taiwan's smart urban systems— whether to control road traffic, manage sewerage, provide healthcare, or monitor security cameras—would fail if they needed to communicate even briefly with a server abroad through an undersea cable. Which is more likely than not: The internet's abstruse routing protocols and network layers, unaware of distance or efficiency or geographic boundaries, often end up shooting packets of information through distant servers rather than local ones. The cloud knows no borders, which means your data may be stored in an entirely different continent—a disaggregation made possible only by the cables at the bottom of the ocean.

A century and three-quarters after the first telegraph line was laid across the Atlantic, there's still nothing to beat the cable. You couldn't fill Earth's orbit with enough satellites to handle the tides of data washing up and down cables every minute. (And satellite companies, in any case, are hardly free of the instincts to profiteer, manipulate the market, and strongarm customers, as the antics of Elon Musk and SpaceX make clear.) Other fledgling ideas, such as bouncing signals from balloon to balloon in the stratosphere or beaming information through the air with lasers, serve brief distances, but they still require the cable on the seabed to work on any great scale. As far as the submarine cable is concerned, there is no Plan B. Within a matter of decades, it has joined the electricity grid, banks, transport links, and the water supply network as crucial infrastructure. It isn't so much an agent of globalization as *the* agent of globalization; without its gift of cheap, instant communication, all the platitudes about dissolving borders and a small, flat world would never have come into being. The safety of the cables in the ocean is a national security issue, a precondition for the economy, and a matter of literal life and death.

Which prompts the question: Who lays these cables, keeps them safe, and makes them whole when an underwater volcano or some other force damages them? And how, for that matter, does one even go about laying a cable in the unseen depths of the sea? In 1996, science-fiction writer Neal Stephenson wrote a rollicking essay for *Wired* on this very subject. By turning up as a self-proclaimed "hacker-tourist" at the nodes and beachheads of internet cables around the world—Tokyo and Hong Kong, Alexandria in Egypt and Penang in Malaysia—Stephenson cast the cable business as a Wild West frontier. He found harsh landscapes, misfits and outcasts, maverick entrepreneurs and witless officials and bullies who didn't seem to realize that their bullying days were nearing their end. (In a key departure from the themes of a John Wayne film, the bullies were big telecom corporations.) The frontier was being remade even as Stephenson roamed it as its balladeer, watching the internet's cables wind and spread around the planet.

But three decades are a long time in the history of the internet, so I set out to anatomize the realm of the cable afresh. There are still misfits and mavericks around, as I found—and witless

28 officials never really go away. But there are also giant companies that didn't exist in Stephenson's time, and that hold enormous influence over where cables go. There are some novel ways of doing things, and many old ways that don't seem as if they will ever change. There are new laws as well as new outlaws. There is an atmosphere of profound geopolitical cynicism, and a mood of mistrust and opportunism—precipitated not least by the US, the nation that sculpts and distorts the internet more than any other—that threatens to tip over into all-out war. Above all, there are plenty of twenty-first-century anxieties—about being left out of the internet, or about who controls the flows of data, or about devious foreign powers eavesdropping on cable traffic or severing cables.

For years, submarine cables existed on the fringes of our consciousness. They lay where no human eyes would ever spot them, and came ashore in remote coastal locations, their presence signaled to those in the know only by a manhole cover set into the soft sand. The people who installed these cables were part of an industry in which everyone seemed to know everyone else, but the industry itself was so small and self-contained as to be practically out of sight. In theory, you could always find out where these cables ran because their routes are compulsively published, so that ships and fishing boats can't plead ignorance after accidentally damaging them. But to do that, you'd have to be aware these cables even exist—and yet they're so reliable and efficient that they allow you to never think very hard about how precisely your TikTok loads or your Netflix streams *Dante's Peak* or your stocks trade. Their very indispensability, though, has now pulled these cables into our fresh attention, and revealed the vulnerabilities of a world in thrall to the internet.

The Cartography of Cables

For the cable-inclined, the pilgrimage always begins in Cornwall, England's long, knobbly toe testing the waters of the Atlantic.

In the Cornish village of Porthcurno, a semicircle of high, rocky ground descends abruptly to a beach that might once have been called secluded, before it was hooked directly up to the whole wide world. The sand is pale and soft, and the water on this early summer day was a bright teal. The paddlers were tentative but determined. The beach itself is, cellularly speaking, a dead spot: Not a bar of phone reception was to be had, however I bent and arched my back. Just above, on the rocks, there is a small Vodafone base station. "I sometimes start my tours by telling people that their mobile phone signals first go to a base station like this one in Porthcurno, and then eventually to a cable," Gareth Parry told me. "It's a light-bulb moment for them. They think it's all done by satellite."

Parry, a short and owlish man, had retired as a professor of physics at Imperial College in London. When I met him, he was chair of the board at PK Porthcurno, a communications museum

30 adjacent to the Vodafone base station. The museum is famous because the beach is famous, and the beach is famous because it was, for perhaps a century, the central node in a telegraphic network that spanned the planet. In 1870, a ship called the *Hibernia* pulled ashore an undersea cable whose other end lay in Carcavelos, in Portugal—the final leg of a system that ran through Italy, Malta, Egypt, the Suez, and Aden, all the way down to Bombay. The first message, sent from Porthcurno to Bombay, read: "How are you all?" A reply returned in five minutes: "All well." If it hadn't happened so fast, this would have been an entirely forgettable exchange.

By the mid-twentieth century, fourteen cables landed in Porthcurno, buried into the soft sand and terminating in a flat-roofed cable station the size of a small shed, which still stands today. The station was locked the afternoon I met Parry there, so we peeped through a window at the metal ducts rising out of the floor, bearing their cables within. Each duct ended in a kind of junction box labeled with the exotic origin at the other end of the line: Carcavelos, Gibraltar, Vigo, and Bilbao in Spain; Brest in France; Faial in the Azores; Saint John's in Newfoundland. The cables made Porthcurno so important that, during World War II, hundreds of tin miners were recruited to dig tunnels into the nearby rock, to house the cable station's equipment and protect it from German bombers. (Who knew better than the British that cables made for valuable targets? In 1914, the day after Britain declared war on Germany, British cable ships belonging to British telegraphic companies were ordered to steam out and cut five German transatlantic cables and six cables running underwater from Germany to Britain.) On Porthcurno beach, the military built a flame barrage: a warren of

underground pipes and pumps that could spew fuel oil onto the
sand, to then be lit into an inferno of fire and smoke that could
deter approaching German invaders.

It wasn't only during years of war that the cable was a projection of power—and, in the British case, of empire. The first undersea cables secured the empire but were also literally made of it. To render them waterproof, manufacturers sheathed them in gutta percha, an "imperishable subaqueous insulating material"—as the Victorians celebrated it—derived from the latex from trees found exclusively in the Malay peninsula. To rule Malaya, as Britain did, was to control gutta percha and therefore the production of cables themselves. In turn, the cables bound the empire together and gave the metropole swift new ways to exert its command over its colonies. By 1907, there were 200,000 miles of cable on the seabed, and around 75 percent of these cables were held by the British. When the empire wished to quell uprisings in erstwhile Rhodesia, or when it needed to be apprised of cotton yields in eastern India, the cables hummed with instructions. In 1929, the British government, keenly aware of how vital its cable networks had become to its colonial enterprise, encouraged a merger of assorted telegraph companies into a giant entity called Cable & Wireless. By the time the flame barrage was installed on the Porthcurno beach, Cable & Wireless operated in nearly 150 locations around the globe. The museum above the beach, in fact, used to be a Cable & Wireless college, where it trained engineers and operators before dispatching them to its offices and signaling stations.

Both the telegraph cable and its telephonic successor, the coaxial cable, bore information along in the form of bursts of electricity. The modern fiber optic cable, first tested as

32 a submarine medium in a Scottish lake in 1979, became the medium of choice for undersea networks. In principle, scientists had already known for decades that there could be no better messenger of information than light, the quickest force in the universe. But it required the invention of the laser—a controlled, high-energy beam of light—and the development of highly purified glass for the fiber optic cable to be born. With both electricity and light, data is encoded in pulses of energy. But even if the speed of electricity in a copper wire were comparable to that of light in glass, the material qualities of the copper—its stubborn resistance and inductance—limit the rate at which electric signals can pulse. The bandwidth of copper, or the volume of data it can carry every second, is smaller. The sharpness of the electric signal in copper also degrades faster than in fiber optics—and that was true even before the advent of the "hollow core," in which light courses through a channel of air trapped in the center of the glass fiber. A factoid on a wall of the Porthcurno museum runs thus: "It takes 0.000006 seconds to send a Harry Potter book along a fiber optic cable. It would take 74 days to send a Harry Potter book along a telegraph cable."

Fiber-optic physics is Gareth Parry's specialty. He can talk for hours about how light races along a strand of glass by bouncing off its inner walls; and about how modulating the amplitude and phase of the light, "like switching a shutter on and off very fast," encodes data into the beam; and about how these marvelous devices became both thinner and more capacious, between the first transatlantic optical fiber carrying 565 million bits a second in 1988 and a Japanese experimental cable carrying 1,000 *trillion* bits a second in 2022. He told me about wavelength division multiplexing, an ingenious technique that mixes many differently colored lasers—each a separate stream of information—into a single beam of light, which then travels through the fiber and is unmixed at the other end back into its various colors, its constituent data streams. In an instant, it multiplies the capacity of a fiber by 10 or 20 or however many colored lasers are passing through it. It was as if a public transport genius had invented an eighteen-decker bus, except that all eighteen decks still somehow fit inside an ordinary bus.

If you overlaid a modern map of submarine fiber optic networks upon an old map of Cable & Wireless cable routes, you'd see astonishingly few differences. The thickest thatch continues to run across the northern Atlantic. A belt of cables from southern Europe dips through the Suez Canal before heading hard east to India. One fiber optic cable from Portugal to Brazil, a diagonal across the Atlantic, is the spitting image of a telegraph cable in a 1901 map. Cables still hug the curves of west Africa in similar fashion, and still land mainly in Mumbai and Chennai on India's otherwise extensive coastline. Part of the reason for the persistence of these routes is simply that the cables embody the same essential structure of geopolitical power: the wealth and might of Europe and the US, the relative poverty of African nations, the Suez as the vital gateway between Asia and Europe. The only significant changes between the two maps are the swollen thickets of fiber optic cables in the waters of the eastern Pacific and southeast Asia, stretching up to the shores of China—emblems of the economic rise of China and its neighbors over the last half-century.

But the other reason so many new cables follow their telegraphic ancestors—such that four data cables still land in

34   Porthcurno—is that the earth is too rugged and unknowable for humans to have their way with it altogether. If a stretch of coastline is rocky or fronted by tall cliffs, it's difficult to bring a cable ashore. Cable companies, like vacationers, look for a beach with soft sand and a gentle slope into calm water; unlike holidaymakers, they'd want this beach to be within spitting distance of the big data centers that will sort and distribute the contents of a cable. In the shallows, a fiber optic cable by itself would be too easily cut by the odd sharp-edged rock, a ship anchor, or a fishing boat trawling the seabed, so it must be routed through the safest waters and protected all the way. As it passes from sea to land, the cable is threaded by divers through a pipe made of metal or stout plastic. Farther out from this landing point, but still close to the coast, the cable is double-armored in steel and, if feasible, interred a couple of meters below the seabed by an ingenious machine called a plow, which digs into the soil with a blade or high-powered jets of water, drops the cable into the trench, and covers it back up.

A few hundred meters off the coast, the pipe is no longer required, but a plow must still bury the armored cable until it reaches a water depth of a mile or so. Ture Jönsson, a Danish cable veteran who helped test the prototype of the modern plow in the mid-1980s, told me that the machine was born out of necessity. "It was becoming a bigger and bigger deal to lose cables that carried a big load of traffic to other countries," he said. Cable layers seek out seabed with the softest sediment possible, he said, "because the cable just sinks in easily, and you could dig it up in a hundred years and it would look just new." If you know the busiest shipping lanes and fishing spots, or where the drifting sea ice is thickest, you try to thread your cable between them—a lesson that was learned in the very dawn of the submarine cable business. The first cable across the English Channel, granted approvals after five years, laid unarmored between Dover and Cap Gris-Nez in 1850, was granted approvals after five years. An inaugural message went across to Louis Napoleon. Almost immediately afterward, the cable failed—likely severed by a fisherman. One enjoyable rumor had it that the fisherman later put a section of the cable on display in Boulogne, as a specimen of rare rubbery seaweed with a core of gold.

Wrangling government approvals to bring a cable ashore is such a complicated task that cable companies have their own permitting departments, staffed with lawyers and geoscientists and fixers and bureaucrat-whisperers. In recent years, as countries aim to preserve the marine ecosystem around their shores, these departments have had to deal with environmental licenses and authorizations as well. Cable-laying companies have had to move a coral reef in Guam, and consider the habitat of frogs in Fiji, and relocate populations of the endangered Morro shoulder-band snail in California. (A manager on the California project tells Nicole Starosielski, in her book *The Undersea Network*, that his boss "was assigned to be a snail watcher, with 'Morro Bay Snail Man' written on his hat.") Increasingly, governments require environmental observers to be on board ships surveying prospective cable routes or putting cable down in shallow waters, so that they can look over the shoulders of engineers and make sure they're following regulations.

In deep water, the fiber optic cable needs no burial or armor. "In fact, if you armored it at that depth, you couldn't recover it to repair it. It would be too heavy," Andy Palmer-Felgate, a submarine cable engineer at a major American tech corporation,

36 told me. Palmer-Felgate was once a windsurfing instructor in Menorca and then worked on an oil survey ship. For the last quarter-century, he has been helping to lay cables around the world, and he tracks cable faults obsessively. In 2022, there were 184 such faults, he told me, and of these, only five or ten occurred in the high seas. "There's an economic advantage to getting the cable out into deep water as soon as possible," Palmer-Felgate said. "You don't need a plow. The cable is skinnier, and the cable ship can pay out the cable ten times faster than it does in shallow water. The cost per kilometer of laying a cable is a tenth or a twentieth of what it is closer to the coast."

But the deeps hold their own perils: the jagged lips of abysses that can abrade a cable over time, or rockfalls down a steep slope, or seismic or volcanic activity. Off the western coast of Africa, a submarine feature called the Congo Canyon runs 175 miles out to sea from the mouth of the Congo River. At its deepest, the canyon's walls cut more than a thousand meters down into the seabed. Since the late 1800s, oceanographer Mike Clare told me, "people laid cables offshore from the Congo River and found that they kept snapping." It turned out that heavy rainfall inland washes masses of sediment down the river and out to sea at such speed that the resulting avalanches cut the cables that run across the Congo Canyon. The canyon is so long, though, that it is almost unavoidable as companies lay subsea cables along the west African coast. Four cables crossing the Congo Canyon broke in August 2023; those had barely been repaired when four other cables broke the following spring in a different submarine trench near Côte d'Ivoire. For cable companies, the best seabed is a boring seabed. Laid cautiously and protected well, a cable can last as long as twenty-five or thirty years—but more often than not, they're rendered obsolete by newer, faster, higher-bandwidth cables well before they reach the end of their lives.

Scrutinizing the ocean floor so closely is a process that consumes both time and money. Survey ships, of the kind that Palmer-Felgate worked on early in his career, scan the seabed with sonar, charting corridors that may be as wide as twelve miles in the deepest water, building relief maps in three dimensions. The maps are so exact, one surveyor told me, that "you can even read the scour marks from moving icebergs, like someone went through the dirt with his fingers." A survey for a long cable route can take many months: The ship moves no faster than a few miles per hour, to get the best possible scan results, and weather events like monsoons or cyclones can interrupt a mission for weeks on end. Plows move even slower—perhaps half a mile every hour. Given all these constraints, cable companies tend to plot their cables so that they run through, and land in, places they already know to be safe. Finding alternatives can be too frustrating or costly. As a result, the surveyor said, "along the historical optimum routes, it's now getting pretty crowded down on the seabed."

One spring week in 2023, I attended a conference hosted by the International Cable Protection Committee (ICPC) in the basement of a Madrid hotel. The ICPC is a loose forum of companies, government bodies, and research institutes—all of them keen to work with each other to best safeguard their precious, expensive cables. The conference, spread over three days, was a cable wonk's delight. Speakers talked about cable repair times, and about how to plan routes "from beach manhole to beach

38 manhole," and about how to deal with a situation when one cable had to cross over another ("Better results in seven simple steps"), and about the legal protection of subsea cables in Uruguay. One Japanese professor talked about his hobby of finding cables out in the wild: snorkeling in shallow water, say, or stumbling upon the round metal lid set atop an old-school beach landing. ("My son, he said to me: 'Daddy, you like manholes so much!'" the professor remarked.) Between panels, the delegates met in the lobby, ate mini-croissants, and talked shop. They hadn't seen each other in nearly three years. During the COVID pandemic, the annual ICPC conference had been canceled, and if you had to meet someone, you did it on Zoom—which, it struck me, was really quite appropriate, given that the conference was about the cables that made Zoom possible in the first place. The attendees were overwhelmingly male; in fact, throughout my research over two years, I met barely any women. In this industry, the nineteenth century's shadow is a long one—and not just in terms of the gender ratio.

On one of the evenings, the ICPC hosted drinks at a bar, and I fell into conversation with Darren Griffiths, a former midshipman in the Royal Navy. Griffiths was the head of marine maintenance at OMS Group, a Malaysian company that lays and repairs submarine data cables. In telling him about my project, I must have said something about wanting to learn how the industry had changed over the years. He laughed and gestured around him. "Yes, but you know—it hasn't changed that much."

He was overstating for effect, perhaps—but not by a lot. The manner in which new cables shadow the routes of older ones is a fine metaphor for the conservative nature of the industry. It's an understandable quality when hundreds of millions of dollars are at stake, and when your main operating environment is the unseen depths of the sea. There's virtue in moving slowly so as not to break things, and also in refraining from fixing what isn't broken. A repair ship today picks a snapped cable off the ocean floor in much the same way as it might have done a hundred years ago: with a grapnel hook dragged back and forth across the seabed until it snags its quarry. Divers still help guide a cable from ship to shore in the final stretch of its landing. Sometimes inertia begets inertia. Cables have landed in Mumbai for so many decades that the city now hosts dozens of data centers. "We've sometimes advised people to land their cables elsewhere, but they're reluctant," one Indian cable executive at the ICPC conference told me. "The reason they give is: If you land it in Cochin, in the south, say, you will in any case have to run the cables back up to the data centers in Mumbai. As a result, you don't really diversify from there."

Until very recently, the industry's major players—a handful of companies, really—all had an old-timey-ness about them as well. They were headquartered in the West or Japan, as had been the case since the beginning of the telegraphic age. One of the four biggest companies that manufactures and lays cables, Alcatel Submarine Networks (ASN), is many generations removed from an ancestor founded in 1898. It still makes repeaters—devices installed every forty to fifty miles on a cable to amplify the signal—in a factory on Enderby's Wharf in Greenwich, the neighborhood where the first transatlantic cable was manufactured in 1857. (As of 2024, ASN is owned by the French government.) Another of the leaders, NEC, started in 1899 in Japan as a manufacturer of telephone switches. A third, the New Jersey–based SubCom, is younger: Its DNA is

40 merely eighty or so years old, and it was once part of a conglomerate that had also purchased AT&T's cable-laying operations. (These days, SubCom is owned by a New York–based investment firm.) Global Marine, a British specialist in installing and repairing subsea cables, traces its lineage back to the company that laid the first cable between England and France in 1850—the one with the core that was supposedly exhibited as gold seaweed. Even the people seem to be lifers, as if tethered to the industry by cables of their own. When I asked Ture Jönsson, the Danish cable expert, about his first job, he told me he'd joined a state-owned telecom company's submarine cable division as a trainee in the winter of 1962. This was 2024, Jönsson was eighty-two and heavily bespectacled, yet he'd just returned from a stint aboard a ship, surveying the sea near Greenland to find a route for a cable on its western shore.

But there have been some shakeups, and the career of Andy Palmer-Felgate, the submarine cable engineer at the American tech giant, has overlapped neatly with nearly all of them. Around the time he was working aboard oil survey ships, in the late 1990s, the first major privately financed cable, stretching from western Europe to Japan, opened for service. Until FLAG, short for Fiber-optic Link Around the Globe, cables had been financed by the various telecom operators along their routes. This seemed to make intuitive sense. These operators were, after all, the users of the bandwidth they were creating, and by virtue of often being state-owned or at least monopolies in their countries, they could easily procure licenses from their governments to land cables. If AT&T planned a cable running from California to Japan and then down to Singapore, say, it would rope in NTT in Japan and Singtel in Singapore, work out a way to split the costs and share the bandwidth, and contract a firm to lay the cable.

But FLAG was privately financed, the money raised from investors. "They broke away from the incumbent, government-controlled telecoms market," Palmer-Felgate told me. As countries along the cable's route deregulated their telecom sectors, FLAG struck deals with tier-two companies to be their local landing partners, rather than giving them stakes in the project. In this way, FLAG ran clear across the world: a first leg from Porthcurno through the Suez, skirting around Sri Lanka, and terminating in Kanagawa in Japan, hitting thirteen countries over its 17,400-mile length; then a second leg across the Atlantic that opened in 2001; and finally a sub-network in east Asia and another in the Persian Gulf. Palmer-Felgate, who had joined ASN in 2001, worked on surveys for the east Asian branch of this mammoth cable system. After FLAG, the private model became the norm, breaking the stranglehold that big nationalized telecom companies had on cables and their bandwidth. "That crashed the price of bandwidth, which went on to become almost an enabler for the internet—because the internet would not be what it is today if it wasn't for the fact that everyone could access it on the cheap," Palmer-Felgate said. "FLAG was a complete game-changer."

Not very long after Palmer-Felgate joined ASN, the dot-com bubble burst, and the market for cables collapsed. "No one was building anything, because they couldn't justify a business case," he said. For a while, ASN had just one cable project on its books, called Apollo: a pair of cables commissioned by Vodafone, linking Cornwall and Brittany with towns in New York and New Jersey. Ordinarily, they'd have needed at least a full summer to

42 complete it, he said. But ASN had just finished building a number of new ships, all named after islands off the coast of Brittany. "They were the most advanced cable ships in the world, and it was like: 'Jeez, we just built these ships for $80 million or $100 million a pop, and now we have no work for them whatsoever,'" Palmer-Felgate said. "So they put all of them onto the Apollo cable." The cable lay was wrapped up in a few months. For some years thereafter, as business cooled, ASN and other cable companies held back from investing in ships and factories. Global Marine sold at least a dozen of its twenty-odd vessels. The manufacturing of cable stock slowed.

Then the cable world's biggest customers cut short these doldrums, crashing noisily through the industry's doors, splashing money about and demanding new cables in a hurry.

In 2007, news emerged that Google had partnered with a few other companies to lay a cable across the Pacific: Unity, running from Los Angeles to a coastal town near Tokyo. Already Google had been building its own servers and running them on bespoke operating systems. Now, recognizing that its business relied on moving data around the world, it had decided to get personally involved in how efficiently the data moved, and how fast. It wasn't ever meant to be a one-off enterprise; in recruitment ads, the company started seeking people who might "be involved in new projects or investments in cable systems that Google may contemplate to extend or grow its backbone." A few years later, Facebook waded in, helping fund the Asia Pacific Gateway, a cable system connecting China to southeast Asia. Then Amazon and Microsoft got in the game as well.

Each of these companies had previously been the equivalent of an anchor tenant in a mall—committing to buying a chunk of capacity in a cable system that someone else was building. "Then these companies thought: 'Why are we using the middleman? Let's cut them out and go direct,'" an executive at one of these big tech firms told me. "It was the logical next step, given their huge data requirements—but also to have control over the operations and maintenance of these cables, which you can't get by buying secondhand capacity over a telephone call." As the sole owner of the cable, you don't have to wait for any partners to complete their interminable legal reviews, or ask their permission to add a branching cable off to a promising new territory. You just do it. And for these tech giants, with their swelling revenues, the cost of, say, a half-billion-dollar transatlantic cable was, if not paltry, certainly a sum they could afford many times over. In the industry, these giants are known as OTTs, short for Over The Top players, and they invest heavily in both partial stakes in new cables as well as full ownership of them. The 2Africa system, which will be the longest operational cable in the world once its 28,000 miles are lit up, was conceived, led, and predominantly funded by Meta. According to one estimate, OTTs have funded more than 80 percent of new cable capacity added in recent years. Their treasuries run so deep they're almost impossible to compete with. They've built new cables so feverishly that some observers now fear that the market is glutted with bandwidth, and that a cable-laying slowdown is in the offing.

What does it mean today for the four big OTTs to be the arbiters of who gets served by the internet and who does not? I've heard it argued that nothing has changed. After governments around the world privatized their telecom industries in the 1990s, the decisions to lay new cables were in any case being made by companies, which were driven purely by profitability

44 considerations. For nearly two decades before the OTTs arrived with their fat checkbooks, the cable industry was, like the rest of the internet, surprisingly unregulated. But this is a different moment, and the OTTs are different creatures. Unlike telecom companies, they've come to dominate not just the infrastructure of the internet but the traffic that flows through it as well—the chats, emails, purchases, documents, video calls, cloud storage, and social media feeds that comprise our waking lives. (According to TeleGeography, the four big tech giants accounted for 69 percent of all international capacity used in 2021.) They monetize not just our access to bandwidth but also our data and our activities. And they do so at a time when, for better or worse, the internet has gone from being a luxury to a necessity—nearly as important to life, livelihoods, and social mobility as education. It's a situation primed for oligopolistic misbehavior, allowing the tech giants to prioritize their own traffic over their cables if they wish, or to hoover up every last scrap of data flowing through them, or to dictate internet policy in the countries they connect. Note, too, the provenance of these four OTTs. The internet's sinews and nerves have never been more American.

The OTTs aside, the other momentous change can be dated, unusually precisely, to 2008, the year that HMN Tech was founded in China. Once owned by the telecom giant Huawei and then spun off, HMN Tech is the first big company in decades to emerge as a competitor to the old coterie of cable-laying firms: SubCom, ASN, and NEC. HMN makes its own cable, in a factory in Dongguan in southeastern China, but until very recently, it relied upon vessels owned by its shareholder Global Marine. "We were literally a rounding error on Huawei's balance sheet," Mike Constable, who was the CEO of HMN Tech from 2014 to 2020,

told me. "But we had two or three of Huawei's most senior executives on our board. So I put two and two together. In my opin-
ion, Huawei saw this as being very strategic to China—the ability
to build these cables." The company grew rapidly; between 2019
and 2023, according to TeleGeography, HMN Tech manufactured
nearly a fifth of all the subsea cables that came online. In 2023,
HMN commissioned its own ships as well.

    "But geopolitics came along and clipped us," Constable said.
He saw it coming as early as 2016 or 2017, as US mistrust of
China's tech sector—and of Huawei in particular—grew. "We'd
bid for projects and not win them, or we wouldn't even be invited
to bid." Constable believes the US leaned on investors in new
cables to avoid using HMN Tech's services. "I mean—forget laying cables that landed in the US, because that was just a pipe
dream, but even cables elsewhere in the world!" In 2019, the
US government blacklisted Huawei; the next year, the Federal
Communications Commission explicitly labeled Huawei a
threat to national security, ostensibly because of the risks that
the company would slip surveillance devices into its telecom
infrastructure. Huawei divested its stake in 2020, but even
subsequently, HMN Tech has remained on the sanctions list,
Constable said.

    The result is a warping of the internet's geometry. The US no
longer allows cables to land on its shores directly from China, for
instance—even though the data traffic between the two countries has never been more voluminous. Instead, cables between
the US and China must interconnect in a third country such
as the Philippines, and then loop up to Hong Kong—a dreadful
detour for the efficiency-minded men and women of the cable
industry. Segments of cable systems that have already been laid

46 now lie unused, because a Chinese company had invested in building them, and because switching those segments on could thus be seen as a violation of US sanctions. For its part, Beijing holds back permits to foreign companies that want to lay cable through the South China Sea, and is constructing its own, parallel network of cables—again, just unnecessary duplication of effort and material from the perspective of industry veterans. A $500 million system called EMA, to be laid by HMN Tech, will run from Hong Kong through southeast Asia, Pakistan, Saudi Arabia, and Egypt before arriving in Europe—a clear rival to a US-backed project called SeaMeWe-6 that traverses the same route. Constable said EMA hints at a growing bifurcation of the internet: a Chinese internet and a Western one, so to speak, the two intersecting but only if required and on neutral soil. In the thirty or so years that we've all been perpetually online, politics has never interfered as heavily with the routing and channeling of our data as it does today.

How to Fix a Cable

On my second day in Tonga, I walked west out of Nuku'alofa—past the dock where cruise ships nuzzled the shore, so outsized next to the beach kiosks that they reminded me of walruses on tiny floes; around the corner of the parliament complex; past the royal palace with its shiny black gates and clay-red roof; and along the coast road. The afternoon was warm, and the sun glinted off the Pacific, so when a police paddy wagon slowed down and offered me a ride, I took it. They let me off outside a small, glass-fronted building facing the sea: the headquarters of Tonga Cable Limited, as well as the landing station of the international cable that came to the country from Fiji. If you look carefully at the road just outside the building, you can spot a scar straight across it: the place where a shallow trench was dug into the tarmac, so that the cable could run underground, from the sea to the shore and right into the blinking, churning machinery of the landing station.

When I first began digging into the cable business, I wondered how I could ever photograph the unphotographable. You

48 can't see the cables at the bottom of the ocean, or the data flowing through those cables. You can, of course, gaze on coils of cables stacked on land—indeed, there were a few stray sections stored higgledy-piggledy in Tonga Cable's backyard, left over from old repairs. But seeing them like this, inert and outside their aquatic element, feels somehow like not quite the real thing. One might as well look at a sausage link to learn about an animal's intestines.

The landing station is a proxy, but a poor one. Every time I entered one, I expected a space that would be almost totemic: a chamber that stood in so perfectly for the vast power and reach of the internet that its very air felt charged and omnipotent. And every time, I was met with air-conditioned disappointment. In the Tonga station, the international cable, having crossed the road and been fed through a couple of insulated ducts under the building's cement yard, emerges from the floor in a utility closet of sorts, offering the first glimpse of itself above-ground. On its black sheath, it bears orange and red labels, reading "CAUTION LASER" and "HIGH VOLTAGE," respectively. Then it is led into the building itself—into a ferociously cold room, where tall stacks of servers and switches sit in cabinets of airbrushed metal. Two companies in Tonga have bought bandwidth on the international cable, and it's in this room that the data traffic is funneled to one or the other customer—either to be dispersed further to people on their phones and laptops on the main island of Tongatapu, or to be diverted into the domestic cable that ends in the northern island of Vava'u. The apparatus whines all day and all night. Next door, in a room so bare that our voices echoed, Tonga Cable stores an AC-DC power converter to keep the international cable electrified and alive, and also a rack

of backup batteries. The cable is subsumed into the servers, and   **49**
every packet of data zips out of the building almost as soon as it
arrives. The cable station's best semiotic value lies in remind-
ing us that, in the online world, the personal is industrial—that
even the most casual Instagram post wouldn't happen without
the heavy-duty equipment undergirding the internet.

Around the planet, cable stations sit on all kinds of coastline: glorious beaches, or the seaward edge of teeming metrop-
olises, or the crevices of fjords, or near forests or deserts. But
the stations themselves are near-identical: universal pieces
of refrigerated internet infrastructure plonked down in dis-
tinctly local surroundings. When they do differ, they differ
in size. Tonga Cable's station is a modest one. The biggest
superstations lease out space to host data centers or servers—
a service called "colocation." They're built to be outwardly
unremarkable but also unbreachable. Just after the September
11 attacks in 2001, US troops moved into position around many
cable stations, and the Department of Homeland Security still
frequently reviews the security of many installations. Often,
these buildings offer no signage or other clues to their pur-
pose. Their specs are hardcore. "Can it take a light plane crash?
It's got a really heavy-duty double-skinned roof. Can it take
an eighty-kilometer-an-hour twenty-ton truck? Yes, it can
because of the way it's been constructed. What if someone
decided to take you out? Can they?" a cable entrepreneur told
Nicole Starosielski. The stations are prepared for fire, flood,
power outages, high heat, frost, and humid days. What will
definitely put them out of action, though, is a cable cut far
out at sea.

————————

The chief executive of Tonga Cable was, at the time of my visit, a dapper, genial man named Semisi Panuve. He'd started his career in the 1980s as an engineer with Tonga Communications Corporation (TCC), a government-owned firm that offered landlines and telegraph services, and then went out to Australia to work for Ericsson. In the early 2010s, as the Southern Cross network was being planned, TCC approached the World Bank and the Asian Development Bank to help fund a branch of it into Tonga: the eventual international cable that started working in 2013. Not long after that, Panuve thought he wanted to retire, and he set up a little café in Fiji. But retirement proved too dull, so he returned to Tonga to run Tonga Cable in February 2020—a month before the country shut down for COVID, and two years before Hunga Tonga-Hunga Ha'apai blew out the cable.

Late on the evening of the eruption, when he thought the ash in the air had drifted out to sea, Panuve set out for the Tonga Cable station on foot. When he was still half a mile away, he saw that the road ahead was blocked with rocks and debris. In some spots, whole boats had been lifted inland. Close to midnight, soldiers arrived to clear the way. Then Panuve, his deputy Sosofate Kolo, and a brace of engineers settled in for a night of work. They checked the servers and power, but nothing was amiss. The alarms on the monitoring system were lit up like a Christmas tree, indicating a cable fault. The diagnosis didn't take very long; indeed, to Kolo, it had been obvious the minute his internet had cut out earlier that evening, in the middle of a Facebook browse. Early the next morning, the team ran an Optical Time Domain Reflection test—sending a series of light signals down the cable and measuring the strength of the

backscattered pulses. That helped gauge where the outage was:
roughly twenty-six miles off the coast of Tongatapu.

This was sufficient information to file an emergency repair request. Like other operators, Tonga Cable was part of a regional cable maintenance consortium, paying a quarterly "subscription" to keep a SubCom ship around in the south Pacific, so that it could attend to cable faults. (Each operator pays a different fee, based on the length of cable they own. Panuve told me it came out to a couple of hundred US dollars per kilometer.) To contact SubCom, Panuve found a satellite phone that Tonga Cable owned. "We had to scramble around to see if it was still working," he said, "because we hadn't used it since 2019, when the cable last broke because of a ship anchor." Then they could do nothing but wait for help to arrive.

Tonga's government, too, was trying to find a way to get back online, and the job fell to MEIDECC, a kind of catchall ministry responsible for meteorology, energy, information, disaster management, environment, climate change, and communications. The volcanic eruption had hamstrung almost all of those portfolios at once—but particularly communication. "We weren't prepared, even though we'd had our cable cut in 2019," Stan Ahio, a ministry official, told me with an embarrassed laugh. The country's two internet providers, TCC and a company called Digicel, owned satellite dishes, which would have been useful in this situation. "But they'd dismantled the dishes, stopped their subscriptions, and put the parts in storage," Ahio said. "We couldn't call overseas." In 2016, Digicel had also installed microwave links across the Tongan archipelago, including a repeater tower on the island of Kao, the highest point in the country. But after 2018, when the domestic cable to Vava'u was laid, no one paid much

52   attention to the microwave network anymore. Out on Kao, the solar panels and batteries powering the repeater started malfunctioning, but Digicel had never sent out a team to repair them, Ahio said. Some of these lapses were understandable; the old satellite links could be expensive for a place like Tonga, running to $20,000 a month just for a patchy speed of 4 megabits per second. But also, even for a country that had experienced a cable break just a few years earlier, it had been so easy to be lulled back into the sense of permanence provided by modern connectivity that Tonga stopped thinking about alternatives and backups altogether.

A day after the eruption, Ahio remembered that MEIDECC owned an old satellite phone. "We'd stopped paying the subscription two years ago, but I thought: 'Everyone outside must know what's happening here. Maybe they'll realize it's an emergency and switch it back on for us,'" he said. His wild surmise about the benevolence of some distant corporate executive proved correct. "The first person I called was my sister, in New Zealand," Ahio said—and here, his chuckles turned into soft sobs at the memory. Nothing here was about cold technology; it was about the proof of being alive. "That was such a difficult time." He permitted himself to cry for a few seconds before regaining his composure. "Then we called the International Telecommunication Union, which is a United Nations agency, to see if we could get some satellite connectivity." It took three or four days to resurrect the satellite dishes on the island, and to get the first meager taste of bandwidth: 120 megabits per second, rationed for the use of ministries and other essential work.

These few satellite-linked locations came to resemble the cybercafes of the 1990s, when internet at home was still a rarity.

In his DHL office, Sam Vea would take in packages, manually fill out dispatch forms, and then drive to the TCC headquarters to send emails to his colleagues overseas over the scant Wi-Fi. Others trickled in as well: importers, civil servants, doctors. The government had to be strict with the internet, Panuve said, and supervise what it was used for. "Like Facebook—that's just a waste of time, right?"

On the outlying islands, there were no satellite phones or dishes waiting to be retrieved from cold storage. Vava'u was a whole island of people trying to tell the world they were all right, and if it hadn't been for Roy Neymen, a sailor who'd berthed his yacht there temporarily, they might have gone many days without getting the word out. On his yacht, Neymen had a Garmin device that sent expensive text messages through satellite, and he used this to contact government agencies in Australia and New Zealand. For a while, he set up a communications center in a local café, where Vava'u's residents could come to dictate their messages to Neymen, as if he were the designated letter-writer in a medieval village, and thus reach their relatives overseas. In two weeks, he sent out 1,600 messages. (At Neymen's request, Garmin covered the costs.)

But this served only a very narrow need. The islanders still had no communication between themselves, and no way to log onto the internet—to send longer emails, or to retrieve money sent to them from overseas, or to conduct their business. The ATMs remained inert, and bank branches couldn't disburse money without first establishing how much their customers had in their accounts. Eventually, when the government began daily flights between Tongatapu and Vava'u, a strange and roundabout solution was improvised. Every morning, a bank headquarters

54   on Tongatapu would download onto a thumb drive a spreadsheet with the account details of Vava'u's residents; the thumb drive would then be flown to Vava'u, where the bank branch could work off the spreadsheet to update withdrawals and deposits; the same evening, the thumb drive would return to Tongatapu, so that revisions could be made to the bank's main database. As an ad-hoc system, this worked reasonably well—until one of the aid missions to Tonga, perhaps from Australia or Japan, brought COVID to the islands. The country locked down in early February, for nearly two weeks. Once again, people had no access to money, or to each other.

A month after the volcanic eruption, SpaceX donated fifty Starlink terminals to Tonga—the country's first fat slice of connectivity. As Tonga's resident DHL agent, Sam Vea has photos on his phone of the momentous day he took delivery—the black cardboard boxes of terminals arriving on a small Air New Zealand cargo truck, shrink-wrapped in plastic and stacked on wooden pallets. If he was beaming, you couldn't tell; even in the afternoon heat, everyone wore masks. The Starlinks were distributed across institutions like ministries and banks but also to public spaces like community halls and restaurants; many went to the outer islands, including Vava'u. "All the Wi-Fi was free, so anyone could come into range and use it," Ahio said. "It was always congested!" People could also buy point-to-point radio links that could throw the Wi-Fi signal a little further, to reach their shops or bakeries or schools. On Vava'u, Brian Meikle gestured around the coffee shop where he and I sat talking—a lovely, airy spot with a pizza oven and a balcony that looked down upon the island's marina. In the summer of 2022, he said, the coffee shop fell within range of a Starlink signal, "and it became an

internet lounge. The reliability wasn't the best, but it was better than nothing. You'd see masses of people come into town just
for the internet." They would sit around for hours, working or
studying, and if the café was full or shut, they'd loiter just outside, pecking at their phones. A foreign diplomat told me that
she'd see Tongans in these Wi-Fi oases late into the night, trying to catch up on work: "One woman told me she was doing an
online course at the University of the South Pacific, and that she
had to sit in her car typing out her coursework." When another
COVID lockdown was imposed in late March, shutting down
even shops and gas stations, people from rural Vava'u often
tried to sneak into town, past the guard posts, to get online. The
Starlinks were simultaneously a blessing and not enough—but
they were Tonga's only bond to the world until the snapped cable
at the bottom of the ocean could be dredged up and repaired.

The modern communications satellite is a remarkable device. A
Starlink costs around a quarter of a million dollars to manufacture, weighs just 500 pounds, and can be carried into orbit a couple of dozen at a time. Elon Musk wants to have 30,000 of them
circling the planet. Others are plotting to spray their own satellites into orbit as well. Amazon has approvals to deploy 3,236
satellites as part of its Project Kuiper. Three Chinese projects—
one of them named "Qianfan," translating into the utterly
romantic "Thousand Sails"—aspire to place a total of 38,000
satellites into space. Even in these teeming numbers, though,
they don't herald any big transformations in how we send and
receive data. A satellite can only offer data transfer speeds of a
dozen gigabytes a second at best—far below the multiple tera-
bytes per second that physical cables transmit. Satellites are

invaluable in emergencies—as in Tonga—or war zones like the Ukrainian front, or for military installations or a ranch in the middle of nowhere. But to serve a community with fast, reliable, and voluminous internet, there's no alternative to a cable.

Not that people haven't tried. The highest-profile attempt to dematerialize the internet—to shoot information through the air—was Loon, a project from Google's X research facility, popularly known as a moonshot factory. X's headquarters, set inside a former mall, is not far from the main Google campus in Mountain View. On a September day, I arrived early for my appointment, so I sat outdoors in blinding California sunshine, watching a Waymo autonomous car—another X project—make slow, endless loops around the parking lot.

In a conference room, I met X's CEO Astro Teller, who wheeled in on a Segway. Teller was born Eric; "Astro" was a high-school nickname, but it proved to be a fine and fortuitous one for the chief of a moonshot factory dedicated to envisioning a polished techno-future. That future included bringing the whole world online. Fiber connectivity is a no-brainer "if everyone lives in cities," Teller told me. "Except, like, three billion people don't live in cities—they live in extra-urban or rural places, or even in groups the size of cities but where the infrastructure isn't there. That seemed like one of the legitimate, huge problems facing the world."

Teller jumped up from his seat and went to the whiteboard to start drawing: a stick figure with a rectangular device in its hand, a circle right above her, another circle at the same level but off to the right. Each circle was a tennis-court-sized balloon made of polyethylene and filled with hydrogen or helium, floating 60,000 feet above the earth's surface—or, to be precise, not

so much floating as riding the currents of stratospheric air, rising and falling deliberately as part of a great, coordinated, cir-
cumplanetary matrix of such balloons. Aboard the balloons were
LTE and cellular radios powered by small solar panels, which
could beam 100 megabits a second of wireless internet down to a
patch of Earth fifty miles in diameter. The balloons also commu-
nicated with each other at one gigabit a second, high-frequency
radio waves bouncing back and forth with the precision of hitting the nose on the head of a moving penny, until they could be
transmitted down to a ground station that was connected to the
internet by—of course—good old-fashioned fiber. There was
no escaping the physical medium, Teller said; Loon depended
on cables—submarine as well as terrestrial—just as much as
any other network.

X launched its first test balloons in 2013 from New Zealand,
and for years, the launches, flights, and retrievals of these balloons were shepherded by a team led by Nick Kohli, who rapidly
acquired the designation of "global balloon concierge." Kohli
used to be an emergency medical technician with a pilot's license
and a background in search-and-rescue, until he got headhunted
by a mysterious team that quizzed him about his knowledge of
radios and asked him questions like: "How would you plan a
multi-day expedition to this set of coordinates?" "It was on Day
One, when I had orientation, that I realized what we were actu-
ally doing," Kohli told me. Through his time at X, Kohli super-
intended the flights of thousands of balloons in tests and pilot
projects. After each balloon reached the end of its lifespan, measured in the hundreds of days, it came back from the stratosphere down to Earth, for Kohli's squad to collect. "We must
have had a 95 percent success rate in collecting these downed

58 balloons," he said. Kohli himself retrieved upward of six hundred balloons—sometimes under the strangest circumstances. Once, X negotiated a soft landing of a balloon with the Colombian government—except that the military didn't get the memo. "You can empathize with them, because they're looking at this balloon, going: 'What is this UFO?'" Kohli said. "There's this picture of two Colombian military guards that had captured our balloon, and they made a big PR moment out of it. They put it in jail! And they put it in a shopfront for people to see that they were protecting the community!"

In principle, and even in pilot experiments, Loon's technology worked. It was a feat just to keep balloons moving in sync, to keep them talking to each other amid the turbulence of the stratosphere, to drape the internet over an area with any kind of stability. You could get a Loon system up and running cheaply and quickly, particularly in emergencies, Teller said, whereas cabling a place that was off the grid would take both time and money. "In Peru, we were able to stay online 75 percent of the time—which, maybe from a telecommunications point of view, is disappointing," Teller said. "But you had to remember that these were remote locations where people's main way of getting email and texts was to give their phones to someone who then drives them into the city." This was not an exaggeration, he assured me: "There was really a big box of phones, and the driver would park somewhere where there was free Wi-Fi, download all the messages, and then drive back." Or someone would build a small hut atop a tree or a water tower, and then rent access to customers who could go up to catch a signal. "Given that, just to have your phone talking to the sky 75 percent of the time was pretty good."

But when Loon eventually shut down, in January 2021, it was for other reasons altogether. To provide connectivity in a country, X had to partner with local mobile service providers, but those companies balked at the fees they'd have to charge customers for Loon to make financial sense, Teller said. The other problem was that international regulations treat balloons more like airplanes than satellites—which meant that X had to negotiate with each government for the legal right to fly through its airspace. "If it's a small country like Cuba, that's no problem— you just go around it," Teller said. "But a really long country like Chile, that was a total nonstarter. You'd have to convince them that you weren't doing anything nefarious, and eventually maybe you'd get a Yes."

With time, Teller believes, X could have solved some of these issues: cut down the drop-offs in signal, reduced the costs, carried more data. In several test runs, Loon used lasers—more focused and high-energy beams compared to the radio signals it usually employed—to carry data, and by setting up a daisy chain of ten or twenty balloons, it could carry many megabytes of information from New York to Cornwall. ("I don't know what our record was, but it was at least 5,000 or 6,000 miles," Teller said—far longer than the transatlantic span between the US and the UK.) The chain would only have stayed unbroken for most of the time at best, though. And while the daisy chain of balloons would have carried just one laser's worth of information, a cable has lasers passing through each of a dozen or more fibers within it—each modulated and multiplexed separately, tuned with the complete refinement that the controlled medium of super-pure, fiber optic glass allows. "There was no way," Teller said, "for us to beat one of those cables."

————

After James Panuve made his emergency call to SubCom to request a cable fix, his first headache was just how much cable was missing. The eruption had cut a fifty-five-mile section out of the middle of the international cable that ran to Tonga from Fiji—an unusually large repair job. Routinely, companies keep extra cable wound in depots located in ports around the world— in Cadiz and Bermuda, in Calais and Catania, in Wujing and Apia. In the Apia depot in Samoa, run by SubCom, there were only eighteen miles or so of spare cable of the sort that could be patched into Tonga's international cable. That was a reasonable length of spare cable to hold for most contingencies—but not for the most powerful underwater volcanic eruption ever recorded. Ordering new cable from a manufacturer would take far too long to resolve this emergency. So, over the next few days, Panuve had to seek and borrow more of the right kind of cable from elsewhere—specifically, from a company in New Caledonia, which also stored its own spare cable in Apia, and could afford to part with thirty-seven miles of it at short notice.

As the company contracted to maintain cables in the south Pacific, SubCom always had a cable ship—a 140-meter-long vessel named *Reliance*, commissioned in the optimistic glow of the years before the 2001 dot-com crash—in the neighborhood. When Tonga's cable broke, the *Reliance* was docked in Papua New Guinea, which was a stroke of fortune; it could have been out at sea, just beginning another elaborate repair. The ship had to wait around for a while for an engineer to fly in from the US— not a straightforward affair in the teeth of the COVID pandemic. Sailing to Samoa to pick up the spare cable and other equipment took the better part of a week, but once the *Reliance* got there,

Tonga Cable's Sosofate Kolo, a senior Tonga Cable engineer,
told me, "Samoa didn't want them to enter the port, because of
COVID. Eventually, after a little pressure from our government,
they let the ship in."

When it embarked on the repair, in waters more than a mile
deep, the *Reliance* found that the rupture in Tonga's international cable was no ordinary matter. Routinely, when a ship drags
its grapnel along the seafloor to snag the snapped end of a cable,
it requires just one or two attempts to nab its quarry. To locate
the eastern break in the cable, off the coast of Tongatapu, the
Reliance required seven drags, even with the help of a submarine
remotely operated vehicle (ROV). Having tethered that end to a
buoy, the *Reliance* steamed toward Fiji to search for the second
break and fasten that to a surface buoy as well—a gnarly task,
because the visibility underwater was so clouded by silt and sediment that the ROV offered few clean images to its operators.
(The landslide triggered by the eruption of Hunga Tonga-Hunga
Ha'apai had not only broken the cable but also pushed it sig-
nificantly farther north than expected.) After finding the second break, the crew meticulously swept the stretch of ocean
between the two breaks, trying to dredge up not only segments
of cable that had been bitten off—with a view to possibly reus-
ing them—but also a repeater worth a quarter of a million dollars. This was tricky business—like fumbling on a wide, dark
beach for a noodle dropped on the sand. The missing sections
had often vanished entirely, or were buried under meters of new
rock and sediment deposited by the landslide. The repeater, in
fact, was never found.

When it had finished interrogating the ocean floor, the
Reliance returned to the cable that ran out from Tongatapu,

62   untied it from its buoy, and spliced it to one end of the spare cable on board. Then, unspooling the spare cable as it slowly headed west, laying it down carefully so that it had plenty of slack, the crew spliced its other end to the cable from Fiji. The splicing is intricate work: first the peeling back of the cable's various protective layers; then the cleaning of the glass fibers in a sonic bath, with what is essentially high-frequency sound, because even the most delicate physical contact with them might shatter them; then the soft placement of the two ends in a fusion splicer the size of a shoebox. An arc of electricity melts the glass fibers and fuses them; if you were anywhere nearby, you'd smell the sharp tang of ozone, as the electricity reacted with the oxygen in the room. Then the glass has to be re-sheathed all over again. The entire process can take the better part of a night, sometimes longer—and it's all done in a room that rises and sinks and sways with the ocean's swells.

All told, the repair of the international cable took five weeks. Panuve told me that Tonga Cable paid a bill of $1.8 million, which included a daily rate paid to SubCom for the use of its ship and crew as well as the cost of spare cable to replace the thirty-seven-odd miles borrowed from the company in New Caledonia. The next, pressing task was to patch the domestic cable to Vava'u—but finding sixty-five miles of spare cable in a depot nearby was out of the question. Panuve had to put in a new manufacture order with ASN, and amid the hectic rush of cable-laying instigated by the OTT players, Tonga Cable and its small-batch request had to wait six months before the factory in Calais got around to them. Meanwhile, not long after the *Reliance* finished its job on the Fiji-Tonga cable, SubCom gave up its maintenance contract for the South Pacific. In all likelihood, an executive with the Southern Cross Cable Network told me, SubCom had decided that, in the OTT boom, it was more profitable to lay new cables than to repair broken ones. (SubCom did not respond to requests for comment.) Tonga's capital—and its ministries, traders, DHL agents, schoolteachers, Facebook merchants, and diplomats—had come back online. Vava'u, though, seemed consigned to an indeterminate future of WhatsApp calls made in public cafés, homework completed in parking lots, and life in general without the internet at hand.

The most striking feature of a cable ship is variously called the drum, the tank, the basket, or the carousel. It is the giant circular container in which the cable is coiled, and from which it is paid out at sea. Usually, it is situated in the middle of the ship, and it looks like a peg atop a LEGO brick—as if another ship with a hole beneath might be fitted neatly onto this one. For a big project, it can take more than a month to load a ship's cable tanks with cable. "The one thing that hasn't changed in 160 years we've been putting cables in the water is that we still hand-pack the cable tank," Mike Constable, the former HMN Tech CEO, told me. "You have engines pushing the cable in, but you still have people walking around on the tank and hand-packing it. You can imagine what a mind-bendingly boring job that is, walking around like that for hours."

When I met him, Constable was a consultant splitting his time between Singapore and Thailand, and he talked with the energy of a preternaturally garrulous New Zealander who had just been sprung from a tight-lipped Chinese corporation. He's an industry veteran to an extreme degree. Just two days after he got his undergraduate degree in marine surveying, he'd flown from New Zealand to Singapore, boarded a ship called the *Pacific Beaver* that was surveying the waters between peninsular Malaysia to Borneo, and thrown up the first of many meals. In mock-defensiveness, he pointed out that the ship was trudging so slowly through the ocean, so as to map its bed, that they ended up sailing for thirty days in the thick of the northeast monsoon. "Even the older guys who'd been offshore a lot were starting to get seasick."

For the first part of his career, Constable served on survey ships—which he described as being just like any other ships, but with acoustic sounders, transponders, and sonar gear bolted on. The stories he tells are not so much about the topography of the ocean floor as about the various allergies of governments to such surveys. He remembers being on one ship that was denied entry into Egyptian waters, and that had to bob about on the high seas just off Egypt while Constable worked his satellite phone to obtain the necessary permissions. "I ran up a bill of £10,000—and this was in 1995!" On another survey, he remembers the Syrian government sending some officials onto the ship, to stay on board while they surveyed the seabed near Lebanon. "Countries do usually send reps when you're near their territorial waters, and these guys didn't tell us who they were," Constable said. "But then a strange thing happened. The only English-language film we had on board was *Pulp Fiction*, and we were watching it every second night. And the Syrian guy sitting next to me pointed to the screen and asked: 'How much does that gun cost?' I had no idea. Then he pulled a pistol out, waved it around, and asked: 'As much as this one?'" The officials, it turned out, were Syrian security personnel, and they were armed even during movie night in the ship's mess. In Taiwanese waters,

Constable said, the authorities attach a special GPS beacon to
survey ships, so that the government can track them all the time,
to make sure they don't stray from their pre-approved working
corridors. Sometimes these sensitivities can halt a ship cold.
Once, while laying a cable through the Bosphorus, Constable's
ship was told to stop and go no farther. "The US military had listening arrays at the top of the Bosphorus, to detect Russian submarine fleets mobilizing out of the Black Sea," he said. The ocean
is vast and mysterious—but, to judge by the amount of military
hardware on the seabed, perhaps not as untouched by human
beings as we think.

During the days Constable worked with ASN, the company
didn't have its own cable ships, so he was often ASN's representative on a ship chartered from another firm. The ship's captain,
Constable said, was still in charge of the vessel itself, "so if I say,
'Go left,' and he says, 'No, that's unsafe,' I have to shut the fuck
up." But Constable would supervise everyone who worked with
the cable itself: the teams operating the plow or operating an
ROV, or testing the cable mid-lay to make certain it hadn't accidentally broken, or joining one long section of cable to the next.

In his relaxed Antipodean way, Constable made the laying
and repairing of cables sound simple. "Here's the beach, right?"
he said, indicating land with a flat palm and a ship with the middle knuckle of his other hand. "You float up close to shore, you
do the cable landing, and then the ship moves off. And then,
thirty or sixty days later, you've come to the end of the cable
on your ship, in which case another vessel with a full carousel
must splice on fresh cable or take it onwards, or you've reached
your destination." If only it were so simple. Landings are uncer-
tain affairs, dictated by the tides, the availability of equipment,

66 and the schedules of dignitaries who insist on being at the beach to be photographed welcoming the cable in. The plowing in the shallows, to bury the armored cable into the seabed, is so slow that the ship can rarely move faster than a mile per hour. In water deeper than 5,000 feet or so, cable layers can let their cable lie unarmored on the ocean floor, but the team onboard has to let out cable to conform to the shape of the seabed—with enough slack that it doesn't ever tauten too much, but not with so much that it lies on the ocean floor in big loops that are liable to become knotted. Today, a Hawaiian company called Makai codes sophisticated software that digests marine surveys and GPS coordinates, so that it can control how and when the cable is paid out. But back in the day, Constable said, "it was just guesstimates, basically." In a way, he added, "the worst thing you can do is go back and look at one that's been laid. Because you know, they'd have been happily doing their thing for ages, and then someone goes out and looks and says: 'Oh, there's a loop in it! That's a contractual issue!'"

A cable lay or repair job can feel dull and endless: repeated scours of the seafloor for the end of a snapped cable, a sluggish sail through deep waters, a lot of waiting around for fine weather. But this is far preferable to the extreme alternative. Nobody wants to be the *Responder*, the South Korean cable-laying ship that caught fire in the East China Sea and sank in September 2020. Closer to the shore, matters get particularly perilous. Constable was once on a cable installation off the coast of Thailand when a diver with an air link to the dive support boat kept complaining that he was feeling ill. By the time his colleagues realized that the air compressor was sucking fumes out of the barge's exhaust, it was too late; the diver died of carbon monoxide poisoning. Divers can get injured by the propellers of boats as they work closer to shore. Squalls can capsize survey boats. The dangers are so manifold that safe tedium is a best-case scenario.

Across the world, there are just sixty-nine or so ships dedicated to laying and fixing cables. Many of these ships are old, and only five new vessels were delivered between 2004 and 2020. It isn't nearly a big enough fleet to handle all the work coming their way—at least, not with the urgency that a ruptured or scanty internet calls for these days. In 2022, ASN had already filled its fleet's calendar for the next two years. It hasn't helped that, for a long time, just four Western companies have owned close to half of all the cable ships in the world: ASN, SubCom, Orange Marine, and Global Marine. But over the past few years, smaller operators in other countries have begun to acquire cable ships of their own. Among them is the OMS Group, headquartered in a skyscraper in southwestern Kuala Lumpur—not very far from the old sites of the gutta percha plantations that yielded cable insulation a century and a half ago.

Within the OMS Group, its executive chairman, a man named Soon Foo Lim, is referred to as "Datuk," a Malaysian honorific bestowed by one of the country's royal families. I imagined Datuk as a tight-lipped and serious magnate—and certainly not as an animated man nearing seventy wearing a T-shirt, shorts, and spectacles with a banana-yellow frame. On the back wall of his cabin hung a huge world map, designed to keep southeast Asia at the center, rather than off to the righthand margin. Next to it was Datuk's shipbroking license—the document that helped make his fortune.

68   When Datuk grew up in the 1960s, in Port Klang, a coastal town in western Malaysia, even a phone was a luxury; he didn't use one until he was fourteen. "Even if my friend stayed three kilometers away, I'd walk to his house to talk to him," he told me. In his early teenage years, he grew fond of the original *Star Trek*, but the only way he could watch it was to hide in the bushes outside the open windows of a rich man's house and watch the television inside. Even if it rained, he stayed put, hooked to the show. He remembered being fascinated with the show's futuristic technology, which included not just teleportation and phasers but also communicators the size of a palm. Now even his grandchildren own such devices, and Datuk offered the requisite complaint for a man of his generation. "All of them, they're now like this," he said, and mimed poring over a mobile phone. I didn't point out that those phones would be useless without the submarine cables that his company and others lay and maintain.

Datuk got his start in the colonial shipping agencies that remained in Malaysia after the country became independent. When he was at C. F. Sharp, a company that manages a ship's affairs and stay in ports, he was asked to assist a ship laying a coaxial cable between Malaysia and Borneo. After that first encounter with the industry, Datuk became a shipbroker himself. He'd help with customs clearance, look after cable engineers who'd come to town, or find crews or provisions. "I was a local fixer," he said. "Whatever the need—crew, permits, urgent logistics— I found a way to make it work. 'You got a pool of shit? Okay, I'll get someone to clean it up.'" In 1983, he founded the company that would go on to become OMS, and began collaborating with NEC, the Japanese telecom and cable company—organizing permits for their ships, staffing them, coordinating their operations throughout southeast Asia. Gradually, OMS began doing the beach side of landings, then near-shore dives and burial, and finally deepwater installations and repairs. Humans evolved from the sea to the land, Richard Sun, OMS's deputy group CEO and Datuk's son-in-law, told me. "We went in the other direction."

In 2017, OMS bought its first cable ship: a vessel called *Ile de Re*, owned and used until then by the French giant ASN, which sold it for $30 million. Since then, cable activity has grown so swiftly, particularly in Asian waters, that OMS has bought four additional ships and plans to swell its fleet eventually to nine. Every time a new cable with greater capacity was laid, Datuk has thought: *This is it. There won't be any more demand after this. I won't have any work next year.* "But surprisingly, the moment the cable went into the ocean—bam! They'd have demand not only for that but for the next two or three cables!" The world was metabolizing data so fast, and there were so few firms capable of laying data cables, that OMS was welcomed into the curious, part-competitive, part-collaborative family of such companies. Kohlberg Kravis Roberts, the US investment firm, committed $400 million to the company in 2023, and a consortium of banks lent it another $290 million in 2024. It was impossible not to spot, in OMS's swelling fortunes, something of a symbol for the twenty-first century—an age during which the West no longer held a monopoly on economic or technological dynamism. Datuk himself was all too aware of this. "Twenty or thirty years ago, the Western companies would have said: 'I'm the only one who knows how to do this,'" David Lim, Datuk's nephew and the company's director of operations, told me. "Datuk wanted to prove that world-class expertise could also emerge from Asia."

70   In March 2023, OMS took over the five-year cable maintenance contract for the South Pacific from SubCom—which meant that, later that summer, when Tonga Cable finally received the cable it ordered from ASN, it was an OMS ship that was pressed into service to fix the domestic cable running to Vavaʻu. The cost of purchasing the cable and paying OMS for the repair ran to nearly $2 million—roughly as much as the international repair, Semisi Panuve, the chief executive of Tonga Cable at the time, told me. "The Australian government assisted us with that—they paid for the majority of it," Panuve told me. "These are big-ticket expenditures for a small country."

The work of resurrecting the cables that make long-distance communication possible is also done, in part, with the help of long-distance communication. The supervisor of OMS's Tonga job, a marine maintenance coordinator named Surahtul Imran Ali, orchestrated the whole assignment without stirring out of Kuala Lumpur. He'd been with OMS for three years, and this was the second repair he'd directed on his own. By the time I met him, he'd done another four—enough to know that every job has its own vexations. In the Solomon Islands, an OMS vessel had to deal with a snapped cable that lay on a reef—which ruled out the use of any grapnel that might accidentally damage the coral. Divers and an ROV had to help recover the cable. (To make things more delicate still, the seabed was cluttered with unexploded ordnance left over from the Second World War. The area had to be surveyed, Ali said, so that the ship faced no risk during the repair operation.) Off Australia, a repair involved a cable system that had been laid in 2002—so long ago, in the industry's time frames, that not many people around today are even familiar with its particular technical flaws and habits. In the best conditions, a repair in the South Pacific can be completed inside seven days, Ali told me. The Australia job took over two weeks.

Of his five projects, Ali found the Tonga assignment the simplest—and even that came with unexpected hiccups. In May 2023, a full year and a half after the volcanic eruption, Tonga's replacement cable arrived at the depot in Apia, shipped from the ASN factory by container. There was enough to patch the sixty-five-mile breakage plus some cable to spare, all in coiled segments of around twelve miles each. An OMS ship, the *Lodbrog*, with a crew of fifty-six, was finishing up an installation in Tahiti. By the time that was finished and the *Lodbrog* had sailed eighteen days to Tahiti, the cable had already been in Apia for a month. "Usually, according to our contract, if a ship is free and it receives a call for a job, it has to mobilize in twenty-four hours," Ali said. "And when the transit duration is small, say three or four days, you don't really have much time to prepare on board." Eighteen days, though, was plenty—although at one point, it looked as if the ship might have to change destination. "While the *Lodbrog* was going from Tahiti to Samoa, we got a call from the Solomon Islands, who really wanted their cable fixed first," Ali said. "From their point of view, I guess they thought: 'The [Tonga domestic] cable has already been gone for sixteen months, surely they can wait.'" OMS leaves customers to discuss the relative urgencies of their repairs between themselves. Tonga Cable refused to defer its repair.

Having picked up the cable in Apia, the *Lodbrog* arrived in Tongan waters on July 3 to begin the work of splicing the new cable into the old. While on its mission to fix the international cable the previous year, SubCom's crew had helpfully located the two ends of the broken domestic cable buried under sediment in

72 the ocean. The *Lodbrog* sailed first to the southern break—the point at which the domestic cable, running out from Tongatapu, had snapped—where it recovered the cable and spliced a segment of new cable onto it. Then, unspooling fresh cable all the while and splicing one segment to the next, the ship inched toward Vava'u, until it met the northern break and made the final connection. The work proceeded around the clock. Siu Moala, a Tonga Cable engineer who was on board the *Lodbrog* as his company's representative, described how, in the middle of the night, whenever OMS's technicians needed him to watch or weigh in on a process, they'd come to his cabin and wake him up. During the day, when the slow, careful lay of the cable was in progress, Moala would go up on deck. "You wouldn't even think the ship was moving," he said.

For each of the five splices they made, the crew used a joining kit, which included two halves of a protective casing to be heat-shrunk into place around the new seam. Here, as Moala told me, the Lodbrog ran into difficulty. Of the ten kits on board, all supplied by ASN, six had casings that bore a scratch on one half or another. The scratches were so tiny the naked eye could barely make them out—and yet the defect caused the casings to warp and split while being heat-shrunk. After all the kits were scrutinized for scratches, OMS's technicians proposed a deft solution: to mix and match casing halves, working only with the ones that were unscratched. It was an issue that OMS had never encountered before, but it worked. Another time, Moala told me, when a newly spliced cable was tested, it transmitted no signals at all. "I said we have to break it," he said, referring to the just-fused glass in the optic fiber, "and join it again." All told, the repair required seven days—exactly as Ali anticipated.

Even after a cable is made whole again, it isn't possible to simply flip a switch and turn the internet on again. A few weeks of testing and configuring must follow: particles of data shimmering up and down until everyone is assured of their speed and their coherence and the quality of their reception. Only in July 2023 did broadband internet return to Vava'u; when I was there, a few months later, people still seemed to be reveling in the novelty of fast and ubiquitous internet. Once again, you could check your email anywhere. Most of the Starlinks terminals were switched off..

The following summer, though, another earthquake wiped out the same domestic cable. Tonga's government had just ordered Starlink to suspend operations in the country, until it obtained a full license, so once again, Vava'u found itself consigned into the dark. When I read about this, I remembered my conversation with Sam Vea, sitting in his office at the Tonga Chamber of Commerce & Industry, the sea breeze filling out his thin crimson curtains. The only way for a cable break to not cripple the country, he told me, was for Tonga to have a second cable—and routed differently, so that an undersea geological convulsion didn't blow them both out at the same time.

Vea handed me a document: an update to the Tongan gov ernment's infrastructure investment plan for the decade, issued after the Hunga Tonga-Hunga Ha'apai eruption. A new domestic cable linking Tongatapu to Vava'u had shot to the top of the government's priority list—above the upgrading of hospitals and water supply systems, above the construction of a new bridge, above the refurbishment of the parliament and courts. The government budgeted its cost at around $17 million. Vea shrugged. Where would Tonga find the money? It was already relying on donors and friendly governments to build its schools

74 and fix its roads, and as a nation of small islands, it also hoped to secure funding under the Paris Agreement to beat back climate change. A new data cable could arrive only as largesse from others. A hundred and sixty-five years after the first telegraph link was laid, these cables still wind around the world just as the great powers and their corporations see fit—to the point that the cables become tokens themselves of the contests for might and wealth, shaping and distorting the internet as we know it.

A Cable Lands

Ever since my fixation with submarine cables began, my white whale was always 2Africa, the 28,000-mile beast that is being looped from Cornwall to Portugal, down Africa's western coast and up its eastern flank, before diverging to head on the one hand to Mediterranean Europe and on the other to the Middle East and India. The cable had been conceived in 2019, on a white-board in Meta's Menlo Park headquarters, and its forty-six landings in countries on three continents will, according to Meta, "deliver seamless international connectivity to approximately 3 billion people, representing 36 percent of the global population." Meta has seven other partners in the 2Africa consortium, but it is bearing the largest share of expenses. No one has officially revealed how much the cable will cost; when I asked one Meta executive if the often-circulated figure of $1 billion was approximately right, he hemmed and hawed and said: "Yeah, it's probably around that."

By virtue of its sheer scale, everything that gratifies us about submarine cables feels amplified a hundredfold in 2Africa; the same goes for everything that worries us. Africa is sorely under-served by the internet. In 2021, only 22 percent of people in sub-Saharan Africa were using mobile internet. The average cost of 1 gigabyte of mobile data was 10.5 percent of a person's monthly income. Only a third of the population across Africa had any kind of broadband at all. Meta claims Africa's economic activity will rise by $37 billion in the two or three years after 2Africa is lit up in 2025. Whether you believe Meta's figure to be self-serving or not, there's little doubt that, in a world so hooked on the internet, it's a handicap for a country to lag behind in its access to the online realm. At the same time, a continent with a dismal history of exploitation by Western corporations is once again compelled by circumstances to rely upon one such corporation. By being the dominant partner in 2Africa, Meta now wields enormous power over the future of communication in many nations. Having spent hundreds of millions of dollars on the cable, Meta will want to prioritize the business of making that money back—by hoovering up user data, serving targeted ads, and pinning browsers within the iron embrace of its various platforms. Countries in Africa, with few data centers of their own, must push their financial and health records, their governance and welfare data, and their security information to data centers offshore—probably through 2Africa to Europe. It's a paradoxical wrinkle in the world of undersea cables. At a time when many wealthy governments are growing increasingly jumpy about their sovereign control over their data infrastructure, poorer nations find that if they want better internet at all, they must relinquish some of *their* sovereign control to Western tech giants.

In Africa, you don't have to search too hard to find a precedent for this. When oil and gas were discovered across the continent, its governments desperately needed the revenues but didn't have the resources to pull these fuels out of the ground. That task had to be ceded to Western oil corporations; even today, they control nearly half of oil and gas production in Africa. The profits from the oil and gas tend to go overwhelmingly to these companies, their shareholders, and their investors. (They also flow to governments—but often as bribes to officials, and too meagerly as tax payments.) It's hard to make the argument, for instance, that Nigeria's abundant oil reserves and more than half a century of extraction have enriched the country when two-thirds of the population lives on less than $2 a day. It's easier to see the corruption and the blight of pollution in the Niger Delta, as well as the imminent, awful consequences in Africa of the climate change caused by burning carbon. The oil industry has been simultaneously indispensable and exploitative—just as the data industry threatens to be. For Africa, oil is indeed the old data.

I first heard about 2Africa in detail from Rick Perry, the head of international network planning at Vodafone, which is one of the partners in the 2Africa consortium. Perry is such an old-school legend in the submarine cable world that Vodafone named its allotted 2Africa fiber pairs "SHARP," short for "System Honoring the Achievements of Rick Perry." "My ex-boss sorted this out," Perry told me. "I think he wanted to stitch me up. It's very embarrassing." Perry, a cheerful man with a kind face and a nimbus of white hair, had been in the cable industry for half a century when we first met. Fresh out of school, he joined Cable & Wireless's training college in Porthcurno, in Cornwall, and after two years of training, he was sent out into

78   the world—to Bahrain, to be precise, where his first job was to monitor the company's international telegraph circuits. As he stayed with Cable & Wireless and moved up its ranks, he got involved in investing in its cable networks. For a long time, this involved buying capacity on other cables, but in the 1990s, his team instigated their first project from scratch: Solas, running between the UK and Ireland. "We came up with it in a pub," Perry said. "All the good ideas come out of pubs."

Back then, anyone planning a cable route had to account for pit-stops: places where the cable could briefly make landfall so that its signal could be re-powered and then sent on its way. It reminded me of airline journeys from the early years of flights, when planes skipped from one airport to the next, taking on fuel at every stop. This was why, for instance, the island of Guam came to host a cottage industry of landing stations: It was the most convenient place for a cable to break journey amid the vast, watery span of the Pacific Ocean. By the time 2Africa came about, though, transmission technology had come so far that the cable could run from South Africa to England on a single dose of power. (Just to be safe, Perry said, 2Africa comes ashore in west Africa anyway, to secure a power backup.) The capacity of cables, of course, has grown as well. 2Africa has sixteen fiber pairs, with an aggregate capacity of 180 terabits per second—a universe away from Solas and its 720 *megabits* per second. In the 1990s, Perry had never imagined that kind of leap, he said. "It's just incredible."

2Africa was an idea even before Meta appeared on the scene. Around 2017, while working with east African telecom operators, Vodafone figured that the best way to cut the price of data was to build a new cable—a "cage system" encircling Africa, Perry

said. The company gathered some partners into a consortium,
including Chinese operators heavily involved in their country's
Belt and Road infrastructure scheme. But the funding didn't add
up. "The project, I think, was going to cost $300 million or so,
and we could never quite get to $300 million," Perry said. "So we
were bumbling along, thinking, 'Well, what do we do now?' And
suddenly Meta told us: 'We're building an African cable system.
Do you want to join?' So we thought: 'They've got the money. Do
we swallow our pride?' We decided we'd join."

Inevitably, planning the industry's longest cable proved to
be one of the industry's heaviest headaches. Landing in dozens of African countries entailed negotiating with just as many
different, and sometimes outdated, regulatory and permitting
regimes. The cable's intended route ran through the neigh-
borhoods of hot wars in Yemen and the Sudan. Off the opposite coast, along western Africa, the cable negotiated the Congo
Canyon by heading farther out to sea and skirting it entirely; the
additional length of cable would be more expensive, but there are
few companies better placed than Meta to absorb that extra cost.
In the first two years of the pandemic, negotiating with bureau-
crats in person was impossible, but getting to them virtually
wasn't always easy either—ironically because of the very bandwidth issues that 2Africa was trying to address. When one South
African team, consisting of scientists and surveyors, sailed out
to chart the seabed, it couldn't return home thanks to pandemic regulations; the ship stayed at sea for three months. Since
employees of Fugro, the surveying company, couldn't jet around
the world as they normally do, the work of scouting coastlines
to pick cable landing points had to be farmed out to local sub-
contractors. In the Seychelles, the beach chosen to land the cable

80 turned out to be a turtle habitat, Perry said. "So we had to have twenty-four-hour watch over a number of weeks to make sure the turtles weren't disturbed."

By the time I met Perry, in late 2023, 2Africa had already branched out and landed in a number of countries. (Five different ships had been needed to lay just the deep-sea part of the cable system off eastern Africa. It took all of 2022.) When I told Perry that I wanted to watch one of 2Africa's landings, he told me I could have my pick: Nigeria, Ghana, and Côte d'Ivoire hadn't been hooked up yet. For weeks, I kept in touch with officials at Bayobab, the African company in charge of landing the cable in these three countries, to coordinate a visit to Ghana. No date had been fixed as yet, one of them kept assuring me—only to have the official call me in late November and say, in very apologetic tones, that the Ghana landing had suddenly happened nearly overnight, because ASN's ship had been in the area and had some time to spare. After the winter holidays, before I could collect myself, the cable landed in Nigeria as well.

That left Côte d'Ivoire. On WhatsApp, I shadowed my Bayobab friends relentlessly, and they reassured me in March 2024 that the landing in Abidjan was imminent. They were just as reassuring in May, and then in July, August, and September. Finally, early in October, the magic message arrived. The cable would land in Abidjan, the old capital, on October 9. If I made haste, a Bayobab executive told me, I could see Côte d'Ivoire be connected to the longest, most advanced submarine cable in the world.

Bayobab's headquarters in Côte d'Ivoire lies just north of Abidjan's long lagoon, in a hilly neighborhood perversely called

Plateau. In the trees lining the avenues, thousands of fruit bats roost through the day and then take flight at dusk. I caught sight of the great chiropteric ascent one evening, as my taxi was driving through the city. Up they swarmed, above the high-rises and treetops, until the skies above Plateau seemed to flicker and fragment, as if a celestial television screen had suddenly filled with snow.

Already, by the time I reached Abidjan on the evening of October 8, a delay of some kind had manifested. The next day, when I visited Florent Guede in his office, he explained it to me. Guede, the managing director of Bayobab in Côte d'Ivoire, is a compact, big-bellied man, and although he laughs often and easily, he had the jittery air of someone in limbo. Côte d'Ivoire had already waited months for 2Africa; ASN's ships had been occupied elsewhere, and the Abidjan landing had been pushed so much that it became the last one on the continent. Now an ASN vessel, the *Ile de Sein*, had sailed to the Gulf of Guinea to complete the landing, but the plow that would trench and bury the cable had developed a snag. Even as we sat in his office, he received an update: The landing would proceed on the morning of Friday, October 11, weather permitting. He beamed and grew less anxious. "Interesting times, interesting times," he said.

Another man sat with us in Guede's cabin, his lean frame folded into an office chair: Ibrahima Ba, a network infrastructure director at Meta. For my purposes, it was like he was both there and not there. We were in Abidjan for the same event, staying in the same hotel, attending the same meetings, and often riding in the same car—but since Meta hadn't permitted me to speak to him, or indeed to anyone else, none of our conversations technically ever took place. Which is a shame, because insofar as any

82   one person is the progenitor of 2Africa, it is Ba—a man who grew up in the north of Mauritania, a country that is almost wholly in the Sahara Desert, and that got some spotty 4G services in its biggest towns only in 2021. Elsewhere, Ba has evangelized the kind of faster, more reliable access to the internet that 2Africa promises. "We believe that connectivity is good for businesses of all sizes, including Meta," he told *The Nation*, a Nigerian publication. "When people come online, we hope that they will consider using our services." The logic was impeccable. To get 1.4 billion Africans to spend money online, it was first necessary to give them cheap, quick data. If the highway toll was too expensive, why would they drive to the supermarket at all?

On Thursday, October 10, with the landing barely twenty-four hours away, I attended a ceremonial meeting during which Bayobab formally told Côte d'Ivoire's telecommunications minister of the cable's imminent arrival. Officials and executives fitted themselves around a long conference room table in a government building, while reporters and camera operators encircled them. For his audience, Guede played up the resilience that 2Africa would provide his country. Until then, four big international cables landed in Abidjan. Three of them, MainOne, ACE, and WACS, had just a year or two to go before they turned fifteen, at which point operators who'd bought capacity on them could choose to leave the consortiums. The fourth, SAT 3, was nearly a quarter-century old—primed to be decommissioned altogether. Not only were these cables on the brink of obsolescence, but they also ran over the Trou sans Fond, a subsea canyon off the Ivorian coast—similar to the Congo Canyon, and posing the same kinds of dangers. Just half a year earlier, in March 2024, a landslide in the Trou sans Fond had ruptured all four cables,

throttling or even blacking out the internet in Côte d'Ivoire and
twelve other countries. Ghana's stock exchange closed late.
Earnings calls were disrupted. Connectivity in Côte d'Ivoire
plunged to 3 percent of ordinary levels. Côte d'Ivoire's internet infrastructure scores a measly 24 percent on the Internet
Resilience Index—lower even than Afghanistan. One estimate
put the country's daily losses from a total outage at $24 million—
and full repairs took the better part of two months.

All this was at the back of Guede's mind as he talked about
2Africa. About how, close to the coast, it would be buried below
the seabed at a depth of two meters. And how it was routed to
avoid the Trou sans Fond altogether. And how it was set up so
that even a local break—in the branch connecting Nigeria to
the main trunk of 2Africa, say—wouldn't affect connectivity
in other countries. The minister nodded along. When Bayobab
invited one of the reporters to ask a question, he wanted to know
only one thing: How much cheaper would mobile data packs
get? The minister grinned, refused to commit himself, and then
stood up for a round of photographs.

That afternoon, Guede took Ba and me to the prospective
landing point. We crawled through cataclysmic traffic to drive
south, past the airport, and then east out of Abidjan and along
the coast. The sea shone in the heat, switching color abruptly
from a light teal to a deeper aquamarine. The shoulders of the
road were lined with billboards advertising varieties of internet
commerce: a gambling app called BetClic, a company that sold
cement "at a click," a grocery delivery service, Uber. All these
businesses were banking on connectivity getting better and better with every passing year—a kind of implicit determinism that
reminded me of other beliefs, such as the one that house prices

84   could only ever go up, or the one that the economy would always grow in the long run. For a moment, I marveled at the small, weird industry of cable-planners and cable-layers who'd turned the internet's proliferation into such an axiomatic fact of life.

In Anani, an exurb a dozen miles east of Abidjan, our cars left the highway and drove through sand and scrub to reach 2Africa's beach manhole. Beneath a door-sized rectangle of raised concrete, Bayobab had sunk a hollow concrete cube into the soil. The connections to the servers in the landing station a mile away had been installed, and a spaghetti of thick black cables lay concealed in the cube below the manhole. All that was needed, when 2Africa arrived on the beach, was for it to be threaded through into the cube, hooked up to the cables, and tested—simple enough to say, but in reality a process that would take weeks. Then Côte d'Ivoire and the rest of Africa had to wait until the spring or early summer of 2025 when 2Africa would be switched on—if everything went according to plan.

Plans, in the realm of submarine cables, are purely contingent exercises. On our way back to the city from Anani, Guede received a phone call: The landing wouldn't happen on Friday, October 10, after all, and there was no telling yet if it could take place on Saturday. Ba and I had both planned to fly out on Friday night, and we spent a fretful twenty-four hours, waiting until Friday afternoon for the news that everything was set for Saturday morning. Thrilled, we pushed our tickets out by a day.

On October 11, we reached the beach manhole just after dawn, to see that the *Ile de Sein* now lay just off the beach, its stern facing us. The curious sight of a large ship so close to shore had drawn no onlookers at all. The only people standing around

were those working on the project—a small, cosmopolitan
crowd of Ivorian telecom officials, a Lebanese contractor with
his two bright-yellow earthmovers, Greek cable technicians, an
American ASN rep, and the Panamanian chief of a diving and
welding company. Ba and I stood on the beach and watched the
distant bustle on board the *Ile de Sein*; then, figuring it would
be a while before the action began, we drove to a nearby gas station for coffee.

When we returned, Guede was standing in a downcast knot
of people. Once again, the landing was off. It turned out that,
as is routine, ASN planned to use a smaller boat to bring the
cable from ship to shore—a fiberglass vessel with twin outboard
engines. That boat had been stationed in the Abidjan port overnight, and when ASN employees went to collect it that morning,
they found it had sunk. "Can you imagine the look on their faces
when they went there and didn't find the boat?" the American
ASN rep said. "That sinking feeling—literally, in this case!" No
one seemed to know how or why that happened. Rustling up a
replacement wouldn't be easy. A cable is usually brought to a
beach tied to orange floats, and an accompanying diver helps
guide it into place and snips away the floats after it has been
laid. The boat therefore needed metal or plastic cages installed
around the engines so that the diver wouldn't get accidentally
cut up by their blades—and finding a boat and then the right
cages to fit onto the boat would take time. "And if we don't start
the job by 8:00 a.m., then it's All Stop until the next day, because
we need all the daylight hours," the Panamanian man, Jonathan
Camacho, told me.

Unable to stay on, I left Côte d'Ivoire that night. When I
called Guede on Tuesday, he sounded weary. The landing hadn't

86   happened on Monday either, he said. They'd found another boat, but when they fired it up and began heading toward the beach, it started sinking too—a coincidence that would have sounded hilariously absurd if it hadn't been borderline dangerous for the men on board. The *Ile de Sein* was still in Ivorian waters, Guede said. But plans had evaporated. He had no idea when 2Africa would finally land in Abidjan.

Cables at War

In "White Noise," a short story by science-fiction writer Garry Kilworth, a British telecommunications engineer has been asked to look into a haunted cable station on the Red Sea. The cable, a telephone link, ran under the sea from Cairo and terminated in a squat building atop a hill, amid a landscape of marshes and reeds: "It whispered of papyrus and reed boats, of infants abandoned in watertight cradles amongst the rushes." The uneasy engineer notices the silence. There is no birdsong, no croak of frogs, no creak of cicadas; there is only white noise pouring out of the station, because someone has left a speaker plugged into the telephone line.

Indoors, he slips on headphones to listen to the white noise (a move that may or may not be out of the technical handbook). He thinks he can make out faint shouts and yells, the rumbling of wheels, and the clashing of metal—perhaps bronze swords on shields? Horses whinnied. Whips cracked. His colleague Ben mentions a recent theory in *Comms Monthly*: that currents of cold water retain old sounds, much like magnetic tape, and that

88   cables can detect these stored impressions. "What we have here,"
he says, "is the flight from Egypt—Moses and his people." Before
the engineer can keep listening, Ben knocks him out, shoves him
out of the cable station, and sets it on fire. They couldn't risk
what might come over the line next, Ben explains: the voice of
God Himself, speaking to Moses. "We mustn't hear it," he says.
"This is not meant for our ears."

The moment humans started sending information over cables,
they also opened the possibility of eavesdropping on that
information. The private was suddenly made both public and
material: electrons streaming down a cable out in the ocean,
promising to reveal—if not the sound of God—the thoughts
and stratagems of friends, enemies, conspirators, and rivals. The
temptation was never going to be resisted. Wiretapping was born
in the same century as the telegraph cable. The British, having
cut German subsea cables during the First World War, inter-
cepted the coded Zimmermann Telegram that was carried out
from Berlin via Stockholm and Washington, DC, to Mexico City
on British telegraph cables. (The leak of the telegram, which proposed a German-Mexican alliance in the event of the US enter-
ing the war, whipped up American appetites to fight.) In 1971,
American divers installed recording devices on a Soviet naval
cable running through the Sea of Okhotsk. Month after month,
divers recovered full tapes and replaced them with blank ones—
an operation that continued for nearly a decade. One American
submarine, the USS *Parche*, was regularly on tapping duties, tar-
geting cables in the Middle East, the Mediterranean, east Asia,
and South America.

Snooping on older coaxial cables involved "reading" the electric impulses coursing through them. With fiber optics, devices
known as "intercept probes" can be installed in cable landing
stations, reading and storing copies of the data as it transits
through the junction. Both the US National Security Agency
(NSA) and the UK Government Communications Headquarters
(GCHQ) have used intercept probes at least as early as 2002,
resorting to what one whistleblower called "vacuum-cleaner
surveillance of all the data crossing the internet—whether that
be people's email, websurfing or any other data." (The whistle-
blower, a former AT&T technician, described the NSA's installation of this equipment in a secret room in his company's San
Francisco premises.) Plenty of other countries listen in on cable
traffic as well. One former SingTel employee told me about a new
cable laid between Singapore and Chennai in 2002, carrying millions of times more data than its predecessor. India's government delayed clearance for the cable, the employee said, until it
was sure its surveillance technology could be upgraded to keep
up. Out at sea, the US has replaced the *Parche* with a submarine
named after Jimmy Carter, which can tap fiber optic cables on
the ocean floor—a task so seemingly straightforward that the
very internet carried by these cables offers assistance for aspir-
ing tappers. On the Fiber Optic Association's antique website,
for instance, a "How to" guide admits: "First of all, tapping fiber
is easy. You can buy optical splitters that plug into the network
like a cable and divert a small amount of the light to a separate
receiver." *PC Pro*, a British magazine, interviewed an expert who
offered another technique. "You can get these little cylindri-
cal devices off eBay for about $1,000. You run the cable around

90 the cylinder, causing a slight bend in cable. It will emit a certain amount of light, one or two decibels," the expert said. The light leak was so minuscule, he added, that "they wouldn't even register someone's tapping into their network."

The majority of internet data flowing through undersea cables today is encrypted, but it can still spill all kinds of beans, as Arturo Filastò, the executive director of the Open Observatory of Network Interference (OONI), told me. OONI, a nonprofit, believes in a free and open internet—which isn't exactly possible if governments are spying on cable traffic. Even if security agencies aren't able to access the encrypted data itself—the contents of an email, or the numbers in a spreadsheet on the cloud, or the amount of money in a bank transfer—they can glean the metadata, the data *about* the data. "And that can be quite juicy by itself," Filastò said.

Filastò, who lives in Milan, was on a video call with me, and he used that as an example. "The video feed is encrypted, but you can determine the fact that we're on Skype, and that it's you and I who are talking, and that the call has lasted nine minutes so far." When an internet browser visits a website, the connection to a server—called "the handshake"—may be unencrypted, Filastò said. "So by sniffing, you can tell if he's accessing Grindr, or Pornhub, or HoorayTheUighursShallBeFree.com." An eavesdropper can learn whom he emails most frequently, how many times he makes Amazon purchases, how long he spends watching Netflix daily, and where he travels. Given that cables transmit so many petabytes of data that they'd be impossible to sift through, it's almost more useful to parse the metadata. A decade ago, Michael Hayden, a former director of the NSA, said: "We kill people based on metadata."

Filastò is remarkably sanguine about the vulnerabilities of the internet. "My personal take is: I think people should stop worrying. Accept that the internet can be tapped, and then work on building protocols that are more resilient," he said. Part of the problem lies in the internet's genesis, as a network to connect universities and not 8 billion humans. Some solutions are evident: a full-blown switch, across the internet, to the IPv6 protocol, for example, which would provide practically limitless labels and addresses to nodes in a network, and therefore an ability to hop between addresses and thus mask one's activity. Easier said than done. "Amazon wanted to go [all in on] IPv6 at one point, but someone in management realized that to upgrade all its infrastructure, it would have to buy the entire world's supply of IPv6 routers for a full year," Filastò said. "At this point, we're really fixing the plane while we're flying it."

As long as undersea cables can betray their data, countries will try to tap the cables of others—and protect their own. Nowhere is that more evident than in the dynamic between the two great powers of our age, the US and China.

In 2012, the US government began moving to restrict the American activities of Huawei and other Chinese telecom companies, on the grounds that they were likely to be exploited by Beijing for "malicious purposes." Primarily, this referred to cyber-espionage: in other words, tapping. These concerns gradually extended to undersea cables. In 2020, via an executive order, President Donald Trump reformed the twenty-three-year-old Committee for the Assessment of Foreign Participation in the United States Telecommunications Services Sector—a mouthful that the industry frequently abbreviates to "Team Telecom."

92 This body, chaired by the US attorney general, influences the granting of licenses to land submarine cables on US soil—but more broadly still, it seems to be determining the route of cables halfway around the world, their investors, and the companies that may or may not lay them.

Even before Trump's executive order, Team Telecom hadn't been shy about using its government's might to stymie Chinese companies. In the early 2010s, a cable system called Hibernia Express was forced to drop China's HMN Tech as its contractor and choose the American-owned SubCom instead—even though the cable only went to the UK, Ireland, and Canada, and never touched American shores. The reviews of applications to land cables became arduous; approvals could take as long as six hundred days. (In Australia, the same process takes just six to eight weeks, I learned.) Mike Constable, the former CEO of HMN Tech, told me that Team Telecom makes US companies sign security agreements about cables that they invest in and that they bring to American shores. These companies don't always share the details of these agreements with other investors in their cable consortium. Team Telecom's review of any cable seeking permission to land in the US is so pettifoggingly thorough that it even demands to know to who supplies the laptops on which the project's work is done, Constable told me. If a cable's primary network operations centre (or NOC) is outside the US, Team Telecom sometimes "also wants a 'secondary NOC' in the US, staffed by people cleared by the US government— basically a big red button on the wall, where if they say, 'We want you to turn that system off,' you push the button."

In the glut of cable construction since 2020, Team Telecom has been kept particularly busy. It recommended that a cable connecting Cuba to the US be nixed. It advised that the Pacific Light Cable Network be denied permission to lay a cable between Hong Kong and the US. After a giant cable system connecting southeast Asia, the Middle East, and western Europe—called SeaMeWe-6—picked HMN Tech as the successful bidder, the US forced the contract to be flipped to SubCom, even though HMN Tech's bid of $500 million was a third cheaper than SubCom's first proposal. To achieve this, Reuters found, the US offered financial grants to multiple telecom firms in the consortium and warned them that working with HMN Tech would risk flouting future sanctions—the carrot as well as the stick.

On at least four other occasions, Team Telecom has staved off HMN Tech's participation in cable projects or rebuffed attempts to cable the US and China directly. On one occasion, a cable project called HKA, laid by Meta and Chinese telecom companies, among others, withdrew its application to land in the US. "They couldn't get authorization, so they just had to cut the cable off at the 200-nautical-mile limit off the US coast," one person with knowledge of the project told me. "It's still like that today." He recalled another cable system, Bay to Bay, funded in part by Meta, Amazon, and China Mobile, having to be reconfigured to run from the US to the Philippines rather than to Hong Kong as originally planned. "Even then, they couldn't get China Mobile out of the consortium, as the US demanded," he said. "The technology has come on since then. That was a six-fiber pair cable. Today they're building twenty-four-fiber pair cables." Bay to Bay is stuck in limbo—designed and commissioned, but in effect already defunct.

Even cable maintenance comes under the purview of Team Telecom. One former British employee of SBSS, another Chinese submarine cable company, told me that if Google, say,

94   experiences a cut to one of its cables in east Asian waters, it would have to turn to the firms contracted to repair cables in that zone. But three companies have that contract at present: one from Japan, one from Korea, and SBSS. "It's gotten to a point where Google or any other American company may not be able to use SBSS to fix their cable," the former employee told me. "But if the other two are busy, who else is going to do it?" Months after I learned all this, I noticed how carefully OMS describes itself on its website: The second adjective it uses, after "global," is "neutral."

When Constable was CEO of HMN, he told me, the World Bank agreed to finance a cable project in the Micronesian islands. Over two years, the Bank accepted and weighed bids from HMN and other cable companies; the scoring process was so transparent, Constable said, that it was clear to him that HMN had the lowest bid. "But then obviously a bit of pressure was applied on the World Bank. They couldn't award it to us, so they had to find a way to disqualify all four companies that had bid for the project," he told me. "It was bullshit. You know as well as I do that Australia or the US would then come out later and donate the money to build the system. I didn't think the World Bank would be party to this. I was shocked."

In apparent response, the Chinese government has engaged its own throttle, delaying permits to any cables planned through the South China Sea—the most efficient passage for systems looking to meet east Asia's appetite for internet connectivity. For one such cable, an OMS executive told me, they received a permit seven years after they applied. Some cables have had to be rerouted entirely, at an expense of millions of dollars, through the waters off Malaysia or Indonesia. Repairs drag on. Typically, under UN convention, companies seek the permission of governments to fix faults only if the job lies within the twelve nautical miles of territorial waters off a country's coast. China demands applications to work anywhere in the South China Sea. "You don't want to go in there without a permit, and get into, for example, a standoff with a navy ship," David Lim, of OMS, told me. "It would become an international incident."

While I was editing this book, an election brought Donald Trump back to the presidency in the US, and if the world had been cynical and full of discord before, it now grew trebly so. In an age of chronic geopolitical strife, it has begun to seem more and more unlikely that submarine cables—so essential, so exposed, so monopolized by the great powers—can remain unaffected. The prospect of the US government's interference in this realm has risen exponentially. These spats are turning the subsea cable industry from a relatively transnational, collaborative club into something more hard-boiled and siloed. SubCom, already one of just a handful of cable-manufacturing and -laying companies, now works almost exclusively for American OTTs and the US government. (The US lays its own military cables under the sea, for communications as well as surveillance, and SubCom is its only contractor. The government pays $10 million a year to have two SubCom ships on call at any time, to respond to a national cable emergency.) Constable recalled a French executive's complaint about being summoned repeatedly by US officials in Paris, to discuss a cable that his company was landing in France, and that was being installed by HMN Tech. Sunil Tagare, a cable entrepreneur who runs a popular newsletter on the industry, points out that all these permitting complications have increased the expense and duration of projects—to the

96   benefit of the US tech giants and the detriment of other compa-
      nies. "I think Team Telecom is clueless about submarine cables
      and trying to create a political fight in this space of the tech
      industry where none existed before," Tagare wrote late in 2024.
      As a result of all these geopolitical constraints, Constable told
      me, "China is basically now building its own network of cables,
      and it's questionable whether Chinese companies will continue
      investing in big cable systems like 2Africa." The infrastructure
      of the internet is at risk of splintering into parallel cable sys-
      tems that use up resources in needless duplications, intersect
      only in neutral locations, and refuse to rely on each other for
      backup. And worse—as the great powers come to regard sub-
      marine cables as "ours" and "theirs," they're much more likely to
      damage each other's cables in acts of truculence or belligerence.
      Even as the internet has grown more indispensable, its infra-
      structure threatens to grow less secure.

      Nangan is a small inkblot of an island, with terrain so hilly it
      makes your calves ache. When you land in its one-room airport,
      after a forty-five-minute flight from Taipei, you could, as I chose
      to do, decide to walk due west. In no particular order, you'd pass
      the village of Jinsha, where they're building a new seawall to
      protect against typhoon-driven storm surges; a base for naval
      frogmen, guarded by a statue of an amphibious man wearing
      both diving gear and boots and carrying a gun; another base, for
      the Fifty-fifth Army, a Taiwanese regiment; a statue of Chiang
      Kai-shek; and another army facility, now abandoned, called the
      Iron Fort. ("The coral rock beside the tunnel is covered in glass
      shards fixed in cement as a defense against the enemy frogmen
      who, in the past, frequently took advantage of the fort's isolation

and its prominent shape to try to land at night and surprise the defenders," a sign reads.) Surprised Taiwanese people will slow down on their scooters as they pass you, wondering who you are. Finally, you will arrive at one of Nangan's westernmost points, climb a flight of stairs to a wood-and-metal pavilion, and arrive at the granite feet of the sea goddess Matsu, holding a ceremonial *hu* tablet and draped in beads. If you follow her serene gaze out over the ocean, you might fancy that you see a smudge of land on the horizon, half a dozen or so miles away. Nangan is one of the Taiwanese outposts nearest to China—nestled so close to the old enemy that the island's hackles are forever raised. It bristles with military nationalism.

In February 2023, half a year or so before I visited, a couple of Chinese ships cut two domestic undersea cables running out from Taiwan's main island. One of these, named Taima No. 3, provided connectivity to Nangan. On paper, the ships were a fishing boat and a cargo vessel, but the Chinese navy so frequently uses ostensibly civilian craft for quasi-military "gray zone" activities that it's impossible to be certain whether the cuts to these cables were accidental or not. Nangan is part of a spray of islets called the Matsu Islands, and cables in this small archipelago had been damaged at least twenty times in the preceding five years. The 2023 cuts, coming on the heels of a year of bitter tension with China, were a fresh reminder of Taiwan's mid-ocean vulnerability. Fifteen international cables connect Taiwan to the world. Its western flank faces China, and its eastern flank is seismically unstable; over a three-year period, Taiwan's international and domestic cables suffered more than fifty cuts as a result of both man-made and natural factors. Were a foreign power to snap those fifteen international cables,

98   Taiwan—the West's buffer against China, and the semiconductor factory to the planet—would be unmoored from the world it needs and the world that needs it.

At the county headquarters in Nangan, I ran into a man named Tsung Chun Yen, who wore glasses, braces, and spiky short hair. The braces might have been new; every time he laughed, he abruptly stopped midway, as if he'd suddenly recalled his metaled teeth. Tsung, who worked in the county's anti-corruption office, grew up near Taipei, and he'd been working in Nangan for four years already when the February 2023 cable cuts occurred. On that particular day, he told me, he'd been in Taipei visiting his family, and when he returned to Nangan and stepped off the plane, he found that his phone didn't even have 2G data. "I felt like I'd entered a different world," he said. Like a hoarder, he'd made sure to download music and television shows before he left Taipei; he'd been rewatching *Friends* on Netflix, he said, and its long seasons sustained him through the coming weeks of internet blackout. The government office had emergency connectivity, via a microwave link, but it was so slow and suddenly overburdened that it took Tsung five minutes to download his email. He was in his mid-thirties when I met him, and he just about remembered the internet access of the 1990s, when his family dialed into the World Wide Web through the telephone line: "The connectivity at the government office was much slower even than that."

Tsung directed me to the Lienchiang County Hospital nearby, where the cavernous lobby was midday-vacant and the receptionist didn't understand what I wanted to talk about. She summoned an ER doctor, who spoke English; he took me to an empty examination room and asked the hospital's IT chief,

Rex Wang, to join us. Wang was in Nikes and cargo pants and a T-shirt with the legend "Feeling Is Living." The ER doctor, who wore scrubs, had only been on the island for two weeks, but he put my questions to Wang and, with some assistance from Google Translate, narrated how the hospital got by in the months after Nangan's cable was severed.

Like the government office, the hospital was left only with patchy microwave access to the internet, Wang said, and its bandwidth was limited. Doctors could run a Google search and sometimes even search for patient records, but if they wanted to upload X-ray files to data servers in Taiwan, that wasn't possible. "In Taiwan, we have a national health insurance scheme, so that if you go from one hospital to another, the images are available to everyone," Wang said. "Also, since this is an island, sometimes we'd need a helicopter from Taiwan to come and take people back for more advanced treatment. But even to apply for the helicopter, you needed the internet." Someone dug out a fax machine, which could ride the emergency microwave link to send and receive sparse patient records to and from Taiwan, but its quality was poor. Through those months, Wang recalled four or five patients requiring urgent transfers to Taiwan—people with strokes or traumatic injuries—so hospital employees copied their records and images onto CDs and flew with the CDs to Taiwan, to apply for the patients to be airlifted out of Nangan. Wang felt it was lucky that the internet hadn't evaporated the previous year, when the COVID-19 outbreak was still rampant, and when every new patient's vaccination record had to be looked up before any treatment was administered. And there were still doctors, drugs, and nurses on hand, so the practice of medicine didn't seize up entirely. Still, for two months, Nangan's hospital was effectively excised from its nation's medical network, while Taiwan's government scrambled to repair the broken cables.

For months before my arrival in Taiwan, I'd been in contact with Herming Chiueh, an electrical engineer who was serving as Taiwan's deputy minister of digital affairs. I'd expected his government to be reluctant to discuss the cable cuts in the Matsu archipelago, but Chiueh was forthright. Taiwan wanted the world to know how precarious its security could be, he told me—and that extended to its essential connectivity. The day before I flew from Taipei to Nangan, Chiueh organized a briefing for me at his ministry, to which he invited an academic, government officials, and an executive from Chunghwa Telecom, Taiwan's largest communications company and the administrator of much of the country's cable network. They were all men in half-sleeve shirts and surgical masks. We sat around a conference room table, and each time I asked a question, the men would talk among themselves first, consult their neatly stapled pages of notes, and then prepare an answer to be conveyed through Chiueh or a translator. I was a supplicant consulting inscrutable gods through a pair of oracles.

On different dates in early February, the two Chinese ships had dropped anchor upon the seabed and then kept going, so that the blades of the anchors likely broke the cables, Chiueh said. "I say this is 'accidental,' and they also said it was 'accidental,' so, 'accidentally,' all this happened within a week," Chiueh said, the scare quotes audible to everyone. Officials knew the waiting time for repairs could run as long as six months; in fact, by the time a Global Marine ship finished the fixes on both cables, it was nearly June. Unlike in Tonga, Taiwan had kept its microwave data links

in working condition. But the speed of this data connection—
shared, at first, only by Nangan's public institutions—was an
inconsistent 2.2 gigabytes a second, less than a quarter of what
the island was accustomed to. Sending a text message could take
up to twenty minutes. "We got complaints," Chiueh said in his
wry way. It took a month to upgrade the microwave relays to a
faster speed, so that the wireless internet could be fanned out
to other residents on the island. The microwave link was vital,
Chiueh said. "We were giving out post-COVID payments of
around $200 to every citizen—and during that period, we were
using the internet to disburse the money."

For Taiwan, a heavily digital island nation, the preoccupa-
tion with protecting its cables—always implicitly from China—
is almost existential. The government is putting into action a
medley of plans to build backups and backups to backups,
Chiueh said. He couldn't share them all with me, but he reeled
off a few. New domestic cables, including another one running
out to Matsu. More landing stations, given that all of Taiwan's
fifteen international cables terminate in just three points on the
main island at present. Better microwave links. Seven hundred
ground-based satellite receivers, to help set up what Chunghwa
executives call "a multi orbit satellite service portfolio." Four
low-orbit satellites, resembling those used by Starlink, to be
developed domestically and launched in 2029, at a cost of nearly
$80 million. Government offices, Chiueh said, were preparing
extreme contingency procedures so that "if, on any day, all the
subsea cables are destroyed, they can still talk to their partners
overseas."

Not long after the cables in the Matsu Islands were cut,
Taiwan's communications authority proposed heavy criminal

penalties for anyone who damaged subsea cables: a fine of up to $3.2 million and life in prison. The law is both harsh and, in the case of foreign actors, essentially meaningless. How would a Taiwanese court even begin to try the Chinese crew of a long-gone fishing vessel? At present, there is no effective, coherent body of law to hold responsible saboteurs of cables at sea. The only guides available are a mess of national regulations and the UN Convention on the Law of the Sea. Jurisdictions overlap furiously: If, out in international waters, a ship flagged in Panama and operated by an Indian crew cuts a cable that lands in several countries along the west African coast and that is co-owned by British, South African, and American companies, who is the perpetrator, who is the victim, and where would a trial take place? The law around undersea cables turns out to be just as murky and uncertain as the submarine depths in which these cables lie.

For more than a century, the positions of cables at sea have been recorded carefully in maps and published—the better to warn ships to avoid them. "But if this data is used the other way, it becomes a vulnerability," Chiueh said. "All countries face this problem now."

Was Taiwan, then, tempted to stop making public the routes of its cables and the locations of its landing stations? I asked.

A conferral took place. Then Chiueh said carefully: "By law, these routes need to be published. Only when they're used for military purposes does the information not need to be announced. But your argument is—it's a very creative way. We are currently evaluating possibilities on how to do that." The vulnerabilities of cables in the twenty-first century—their sudden elevation into prime targets—could roll back decades

of transparency. On the other hand, Chiueh said, "if you have a
cable that isn't on the map, in general it will be cut more often—
and those cuts will really be accidental, because ships can't avoid
them. Here is the dilemma."

Just before I left the ministry, Chiueh told me a curious
story. At that moment, of the fifteen international cables landing
in Taiwan, only one—called TSE 1—was connected to China. It
was inoperative; some fault had plagued the cable for more than
a year, and Chiueh said neither China nor Taiwan had shown any
intention of repairing it. But the two countries still exchanged
plenty of digital information, so these data packets had to
travel thousands of miles via Japan or the Philippines, rather
than across the 110-mile Taiwan Strait. "Right now, Taiwan is
physically closer to China, but on the internet, we are closer to
Japan or the US," Chiueh said. Like a wormhole, the submarine
cable had bent time and space until geography made no sense
anymore.

A month or so after I returned from Côte d'Ivoire, 2Africa
arrived in Abidjan after all. An engineer on the project sent me a
short video of the landing: the cable tethered to a series of bright
orange buoys, being winched slowly and carefully through the
waves and onto the shore. The lead buoy was, as always, a ceremonial one, emblazoned with the words "ABIDJAN LANDING"
and the logos of all the partners in 2Africa. The sand shone in the
sun. Men in fluorescent jumpsuits and blue helmets urged each
other to go slowly. Within a day, the cable was buried and covered up. If you'd visited the following morning, you'd have perhaps noticed that a narrow strip of beach was tidier than usual,
but there was little else to give the site away.

There was still work to be done on 2Africa—lighting up the full loop around the continent, building out its eastern arm to Asia—but already rumors were fizzing that Meta had plans to beat its own record: a longest-ever-cable to top its longest-ever-cable. It would be laid in the shape of a cross-planetary "W," the cable expert Sunil Tagare, who broke the news, wrote: zigging from the US eastern seaboard to South Africa, zagging up to India, veering southeast to Australia, and then crossing the Pacific to California. Cleverly, the route avoided the common choke points of cable networks: the Strait of Gibraltar, the Red Sea, Egypt, the South China Sea. Tagare reckoned the cable cost as much as $10 billion to build, lay, and operate. He called it "one hell of a cable."

A few months later, Meta officially announced the cable, to be called Waterworth and to span 31,000 miles. The company didn't offer a cost estimate or many other details, but we can already be certain of some things. If Meta owns the full capacity on "W," rather than bringing on partners as it did with 2Africa, it will mark an escalation in Big Tech's brute-force, winner-take-all approach to cable ownership. Google will doubtless consider a rejoinder—a circumnavigatory mega-cable of its own to add to the industry's arms race. Neither of these cables is likely to sell capacity to Chinese or Russian companies, requiring those countries to lay bigger cables of their own. Not very long ago, it made little sense to speak of "American cables" or "Chinese cables." But as Meta and Google engirdle the earth in cables, and as governments panic about security, the corporate and nationalist constrictions of the internet will align and overlap. If the infrastructure splinters into large, disparate cable systems that meet only if necessary, it will become impossible to think of the internet as a single resource. Far from unifying the world, a fragmented internet will only broaden the gulfs between people and countries. Maybe people won't mind. When Neal Stephenson wrote "Mother Earth, Motherboard" in 1996, he seemed to accept the mood of the time that a smaller, more open world—a globalized world—was an indisputably good thing. The present mood, circa 2025, suggests clarity, too, but in the opposite direction.

The history of essential transmission systems can only ever be a partial guide to the future of the undersea cable. Private companies built the first railroads, telegraph lines, and telephone networks. In many countries, once these systems grew sufficiently critical to security and the economy, they were nationalized; elsewhere, they stayed in private hands even if they were regulated and, more than occasionally, manipulated. But never have these companies been as transnational, wealthy, and concentrated as the tech conglomerates are today. Also, these previous systems were not conduits for simply *everything* the way submarine cables are: our love letters and prescriptions, our calendars and calculations, our gifts and bequests, our library requests, our missile codes and stock picks, our call centers and Netflix, our wages and thefts, our working weeks and Sunday rests. The data is, in theory, ours, but we determine neither where it resides nor which cables it traverses, and we have practically no alternatives to these digital pathways—all of which amounts to an absence of control that was not as stark with older technologies. The submarine cables present challenges that are without any true precedent, so they require new solutions to corporate grasping and international pugilism if the internet is to be secure, useful, and open to everyone.

During my trip to Kuala Lumpur, Wahab Jumrah, a bright-eyed manager at OMS, drove me to one of the company's cable stations in Morib, a beach town on the western shores of the Malaysian peninsula. Morib is one of the landing spots for MIST, a 5,000-mile system connecting Myanmar, Malaysia, India, Singapore, and Thailand. We spent longer than was strictly necessary among the servers in the cable station, in its delicious, frigid air, before wandering out to look at the square beach manhole set into the heat-baked earth. The cable had only just landed three weeks earlier. The facility still had that metallic new-cable-station-smell of fresh paint and recently unboxed electronics.

After the tour, we drove to a nearby resort for lunch. The water park on-site had shut down during the pandemic, and it was now rumored to have been bought by Chinese investors. A few kiosks still sold food, and we brought our fried chicken, rice, and lemonade to one of several deserted tables on the promenade. Jumrah told me about his job, which involved wrangling government permits for cable routes but also cajoling Morib's fishermen to keep away from the thin marine corridor that brought the cable ashore. "We had to figure out how much they earn, we had to meet with the chiefs of these villages, we had to come to an agreement on how to compensate the fishermen," he said. "I kept coming back here again and again for three months until they were happy." Listening to him, it struck me how misleadingly easy it is to regard the internet as a place that is wholly online, wholly abstract—and to believe, therefore, that improving the internet is only a matter of building better websites, better security protocols, better streaming and payment gateways. In reality, though, it is also a matter of negotiating with

fishermen, finding space for cables in waters congested with
ships, purifying glass, charting underwater volcanoes, pacifying
an unfriendly neighbor, and a thousand other tasks that are both
very material and very local, rooted in physical places with people of flesh and blood. I realized this on the edges of coastlines,
by the Pacific in Tonga, the South China Sea in Taiwan, or the
Gulf of Guinea in Côte d'Ivoire. And I realized it again in Morib,
gazing over Jumrah's shoulder at the Strait of Malacca. Beyond
the sandy flats, the ocean was a pale blue wash below the bright
white sky, boats above and cables below, the liquid medium of
the modern world.

FURTHER READING

Stephenson's "Mother Earth, Motherboard," published in *WIRED* in December 1996: https://www.wired.com/1996/12/ffglass/.

And the best academic text on the subject—elegant, thoughtful, and adventurous—is Nicole Starosielski's *The Undersea Network*, published by Duke University Press in February 2015.

Tung-Hui Hu's *A Prehistory of the Cloud*, published by the MIT Press in 2015, is a terrific exploration of the ideas and infrastructure underlying that nebulous concept known as "the cloud." Andrew Blum's *Tubes*, published in the UK by Penguin Books in 2013, is also an indispensable guide to the hardware of the modern Internet.

In *The Cable: The Wire That Changed the World*, published by Tempus in 2003, Gillian Cookson offers a concise and lively history of the first submarine telegraph cable to be laid across the Atlantic. I found a wealth of technical information in *Cableships and Submarine Cables*, by K. R. Haigh, published by Adlard Coles in 1948, and in C. E. Roden's *Submarine Cable Mechanics and Recommended Laying Procedures*, published by Bell Laboratories in 1974. *Communications Under the Seas*, edited by Bernard Finn and Daqing Yang, and published by MIT Press in 2009, is an excellent collection of essays by experts.

Harriet Crawley's *The Translator* (Bitter Lemon Press, 2023) is perhaps the most enjoyable of the recent thrillers to use vulnerable undersea cables as a plot device. A more literary treatment is available in *Twist*, by Colum McCann (Bloomsbury, 2025), set predominantly on a cable repair vessel off western Africa.

ACKNOWLEDGMENTS

The idea for this book came from my friend and longtime *Guardian* Long Read editor David Wolf, whose fertile imagination for journalistic projects I hope to draw upon throughout my career. Surabhi Ranganathan offered further encouragement and ideas during a boozy evening in Brixton; this couldn't have been written without her.

The International Cable Protection Committee (ICPC) and its executives were warm and receptive to this project and generously invited me to their plenary in Madrid. That single event opened so many doors and led me down so many routes of research. Apart from the ICPC office-bearers quoted in this text, I am grateful to Ryan Wopschall, Dean Veverka, and John Wrottesley for their assistance.

Various friends, advisors, and well-wishers guided me or lent me the benefit of their expertise: Franklin Adzimah, Jim Cowie, Mitch Cheney, Ndi Mokom Desmond, Georgina Asare Fiagbenu, Fredrik Galtung, John Gravois, Florent Guede, Abdul Wahab bin Jumrah, Elhad Kassim, Steffen Knodt, James Liu, Nitin Pai, Srinath Raghavan, Audrey Tang, David Tossell, Motohiro Tsuchiya, and Isobel Yeo. Many thanks to Sunil Amrith, who very kindly invited me to present some of this research as a Walker Lecture at Yale's MacMillan Center for International and Area Studies.

It's been a pleasure to work with Nicholas Lemann and Jimmy So at Columbia Global Reports, and I'm indebted to Sujay Kumar for his patient and thorough fact-checks.

My deepest devotion to Shruti Debi at the Debi Agency and Anna Stein at CAA, for helping this slim book stagger out into the world; to Shruti, Varun Rajiv, and Padmaparna Ghosh, my perennial first-readers; and to Vidya, K. R. S., and Harini, for their unstinting support. And, as ever, all my love to Padma, who makes our life possible in so many ways and enriches it beyond measure with her curiosity and spirit.

NOTES

CHAPTER ONE

15 **these islets were among the** *makafonua*, **or land-stones, of Hikuleo:** Edward Winslow Gifford, *Tongan Myths and Tales* (1924) (Kraus Reprint, 1971), 15, https://archive.org/details /tonganmythstaleso000giff /page/14/mode/2up?q=hunga+tong.

15 **2.4 cubic miles of sediment and molten rock:** 10 cubic kms, converted: Jonathan Amos, "Tonga Volcano Eruption Continues to Astonish," BBC, December 13, 2022, https://www.bbc.co.uk/news /science-environment-63953531.

15 **at least thirty-five miles up into the atmosphere and possibly higher:** Stephen Vicinanza, "Scientists Say Tonga Eruption Hit with the Force of a Billion Ton 'Magma Hammer,'" *Interesting Engineering*, updated March 12, 2025, https:// interestingengineering.com /science/scientists-say-tonga -eruption-hit-with-the-force-of -a-billion-ton-magma-hammer.

15 **outdoing any nuclear bomb ever detonated:** J. S. Díaz and S.E. Rigby, "Energetic Output of the 2022 Hunga Tonga–Hunga Ha'apai Volcanic Eruption from Pressure Measurements," *Shock Waves* 32 (2022): 553–561, https://doi.org /10.1007/s00193-022-01092-4.

15 **They heard the sound in Alaska:** Rashah McChesney and Maggie Nelson, "The Eruption Near Tonga Was So Powerful You Could Hear It in Alaska," Alaska Public Media, January 17, 2022, https:// alaskapublic.org/2022/01/17/the -eruption-near-tonga-was-so -powerful-you-could-hear-it-in -alaska/.

15 **in the south Indian city of Chennai, meteorologists measured:** "Shockwave from Tonga Volcanic Eruption Recorded in Chennai," *The Hindu*, updated January 17, 2022, https://www.thehindu .com/news/cities/chennai /shockwave-from-tonga-volcanic -eruption-recorded-in-chennai /article38281020.ece.

16 **through which light transmits information at roughly:** F. Poletti, N. V. Wheeler, M. N. Petrovich, Naveen Baddela, Eric Fokoua Numkam, J. R. Hayes, D. R. Gray, Zhy Li, Radan Slavik, and D. J. Richardson, "Towards High-Capacity Fiber-Optic Communications at the Speed of Light in Vacuum," *Nature Photonics* 7 (2013): 279–284, https:// doi.10.1038/nphoton.2013.45.

16 **then a final sheath of nylon soaked in tar:** "Submarine Cable Frequently Asked Questions," TeleGeography, n.d., https://www2. telegeography.com

/submarine-cable-faqs-frequently
-asked-questions.

17   conducting 99 percent of all
the world's internet traffic: "Fibre
Optics: The Journey Through
Undersea Cables," Telefónica, n.d.,
https://www.telefonica.com/en
/communication-room/blog/fibre
-optics-the-journey-through
-undersea-cables/.

17   have laid 870,000 miles of fiber
optic cables under the ocean: Olivia
Solon and Mark Bergen, "Fishing
Boats Can't Stop Running Over
Undersea Cables," *Bloomberg*,
April 4, 2022, https://www
.bloomberg.com/news/articles
/2023-04-24/fishing-boats-keep
-running-over-ocean-internet
-cables.

17   Cables set out from places like
Crescent Beach in Rhode Island:
All details from the Submarine
Cable Map, https://www
.submarinecablemap.com/.

18   A Finnish company: Commercial
disputes caused the project to
be suspended. But another cable
venture called Polar Connect,
charted to run right through the
North Pole, has taken its place.

18   the cable was designed to shave
twenty to sixty milliseconds off the
speed of trades: Tung-Hui Hu, *A
Prehistory of the Cloud* (MIT Press,
2016), 4.

18   Antarctica is the only major
uncabled landmass on Earth, but it
won't be for long: Dan Swinhoe, "US
NSF Publishes Study on Potential
Subsea Fiber Cable to Antarctica,"
DCD, January 3, 2024, https://www
.datacenterdynamics.com/en/news
/us-nsf-publishes-study-on
-potential-subsea-fiber-cable-to
-antarctica/.

18   cable that connects Tongatapu to
Fiji and thence to the world: "Tonga
Cable," Submarine Cable Map,
https://www.submarinecablemap
.com/submarine-cable/tonga
-cable.

23   to a time before the telegraph and
scheduled flights: *Colonial Reports—
Annual* (Great Britain, Colonial
Office, 1914), 12, https://www
.google.co.uk/books/edition
/Colonial_Reports_annual/8
NWs6fEdJrEC?hl=en&gbpv
=1&dq=tonga+telegraph&pg
=RA30-PA11&printsec=frontcover.

24   Swedish investigators boarded
one such ship: Reid Standish, "Baltic
Sea Incidents Put Spotlight on
Russia's 'Shadow' Fleet," Radio Free
Europe, January 29, 2025, https://
www.rferl.org/a/baltic-sea
-sabotage-undersea-cables
-russian-shadow-fleet/33290689
.html.

24   maritime activities of Houthi
militants in Yemen: Olivia Solon
and Mohammed Hatem, "Houthi-
Sunk Ship's Anchor Likely Severed
Sea Internet Cables," *Bloomberg*,

112   March 6, 2024, https://www
.bloomberg.com/news/articles
/2024-03-06/anchor-from
-houthi-sunk-ship-likely
-damaged-undersea-cables.

**25   NATO launched a mission named
"Baltic Sentry":** Laura Gozzi, "Nato
Launches New Mission to Protect
Crucial Undersea Cables," BBC,
January 14, 2025, https://www.bbc
.co.uk/news/articles
/c4gx74d06ywo.

**25   UK now has warships prowling
its waters to protect its cables:**
"Analysis: Royal Navy Deploys
Seven Ships on Underwater
Infrastructure Patrols," *Navy
Lookout*, December 3, 2023, https://
www.navylookout.com
/analysis-royal-navy-deploys
-seven-ships-on-underwater
-infrastructure-patrols/.

**25   fifteen submarine cables tie
Taiwan to the rest of the world:**
"Taiwan," Submarine Cable
Networks, n.d., https://www
.submarinenetworks.com/en
/stations/asia/taiwan.

**25   an earthquake ruptured
eleven cables traversing an
undersea canyon:** "Cable
Breaks Following the Pingtung
Earthquake," ResearchGate, June
2014, https://www.researchgate
.net/figure/Cable-breaks
-following-a-the-2006-Pingtung
-earthquake-epicenters-of-main
-and_fig4_263506188.

**25   cables carry $10 trillion in
financial transactions daily:**
Tim Stronge, "Do $10 Trillion of
Financial Transactions Flow Over
Submarine Cables Each Day?"
Teleography blog, April 6, 2023,
https://blog.telegeography.com
/2023-mythbusting-part-1.

**25   rely on the cloud just as
companies elsewhere do:** "Cloud
Alliance," TSMC, n.d., https://www
.tsmc.com/english/dedicated
Foundry/oip/cloud_alliance.

**26   through which 48 million
passengers pass every year:** "2019
Airport Traffic Report," Port
Authority of NY and NJ, May 18,
2020, https://www.panynj.gov
/content/dam/airports/statistics
/statistics-general-info/annual-atr
/ATR2019.pdf.

**26   air traffic controllers often speak
to airplanes and to each other over the
internet:** "CERTIUM VCS -
Voice System for Air Traffic
Control," Rohde and Schwarz, n.d.,
https://www.rohde-schwarz.com
/products/aerospace-defense
-security/voice-systems/certium
-vcs-voice-system-for-air-traffic
-control_63493-804354.html.

26   to control road traffic: "Taiwan
Traffic Monitoring Systems Utilize
SSI RFID Readers and Antennas,"
Star Systems, n.d., https://star-int
.net/taiwan-traffic-monitoring
-systems-utilize-star-rfid-readers
-and-antennas/.

26   **manage sewerage:** "Intelligent Cloud Sewer Management System," Stantec, n.d., https://www.stantec .com/en/projects/taiwan-projects /intelligent-cloud-sewer -management-system.

26   **provide healthcare:** https:// www.tandfonline.com/doi/full /10.1080/23288604.2024.2375433 #abstract.

26   **monitor security cameras:** "Taiwan Expands Smart City Infrastructure," SmartCities World, August 14, 2023, https://www .smartcitiesworld.net/citizen -security/citizen-security/taiwan -expands-smart-city- infrastructure.

CHAPTER TWO

30   **The first message, sent from Porthcurno to Bombay:** "Documents Reveal Contents of the First Telegraph Message Between India & England," *Economic Times*, updated February 22, 2012, https:// economictimes.indiatimes.com /documents-reveal-contents-of -the-first-telegraph-message -between-india-england /articleshow/11976488.cms ?from=mdr.

30   **the day after Britain declared war on Germany:** Gordon Corera, "How Britain Pioneered Cable- Cutting in World War One," BBC, December 15, 2017, https://www .bbc.co.uk/news/world-europe -42367551.

30   **were ordered to steam out and cut five German transatlantic cables:** "The Zimmerman Telegram," The National World War I Museum and Memorial, n.d., https://www .theworldwar.org/learn/about-wwi /zimmermann-telegram.

30   **the military built a flame barrage:** Bill Glover, "Porthcurno Cable Station," History of the Atlantic Cable and Undersea Communications, n.d., https:// atlantic-cable.com/CableCos /Porthcurno/index.htm.

31   **"imperishable subaqueous insulating material":** Gillian Cookson, *The Cable: The Wire That Changed the World* (Tempus, 2007), 45.

31   **By 1907, there were 200,000 miles of cable on the seabed:** John Tully, "A Victorian Ecological Disaster: Imperialism, the Telegraph, and Gutta-Percha," *Journal of World History* 20, no. 4 (2009): 559–579, https://www.jstor.org/stable /40542850?mag=the-colonial -history-of-the-telegraph&seq=10.

31   **around 75 percent of these cables were held by the British:** Stephen Luscombe, "Telegraphy: Spanning the Continents," The British Empire, n.d., https://www .britishempire.co.uk/science /communications/telegraph.htm.

31   **When the empire wished to quell uprisings in erstwhile Rhodesia:** Ian Tawanda Mugowa, "Telegraph

114   in Zimbabwe: A Tool of Imperialism," Porthcurno, n.d., https://pkporthcurno.com/pk150/telegraph-in-zimbabwe-a-tool-of-imperialism/.

31   encouraged a merger of assorted telegraph companies into a giant entity: Steward Ash and Bill Burns, "The History of Cable and Wireless," Subtel Forum, July 26, 2022, https://subtelforum.com/stf-mag-feature-the-history-of-cable-wireless/.

31   first tested as a submarine medium in a Scottish lake: ICPC history, 15.

32   the material qualities of the copper: "Why Is Fibre Optic Technology 'Faster' Than Copper?" ABC Science, October 21, 2010, https://www.abc.net.au/science/articles/2010/10/21/3044463.htm; "So electricity travels at 99% the speed of light through copper and fiber travels at 70% the speed of light. Why is fiber 'faster' in terms of latency?" Reddit thread, https://www.reddit.com/r/Home Networking/comments/1cls1jm/so_electricity_travels_at_99_the_speed_of_light/.

32   the first transatlantic optical fiber carrying 565 million bits a second in 1988: Jeff Hecht, "Fiber Optic History," JeffHecht.com, April 2016, https://www.jeffhecht.com/history.html.

32   a Japanese experimental cable carrying 1,000 *trillion* bits a second: "World's First Successful Transmission of 1 Petabit per Second in a Standard Cladding Diameter Multi-core Fiber," NICT, May 30, 2022, https://www.nict.go.jp/en/press/2022/05/30-1.html.

34   interred a couple of meters below the seabed: "MMO Subsea Cables Desknote (2021)," https://www.escaeu.org/download/?Id=390.

35   inaugural message went across to Louis Napoleon: John Watkins Brett, "On the Submarine Telegraph," History of the Atlantic Cable and Undersea Communications, March 20, 1857, https://atlantic-cable.com/Article/Brett/index.htm.

35   rumor had it that the fisherman later put a section of the cable on display: Bill Glover and Bill Burns, "The Submarine Telegraph Company," History of the Atlantic Cable and Undersea Communications, n.d., https://atlantic-cable.com/stamps/Cableships/indexstc.htm.

35   had to move a coral reef in Guam: Nicole Starosielski, *The Undersea Network* (Duke University Press, 2015), 147–148.

[35]   **governments require environmental observers to be on board ships surveying prospective cable routes:** https://engineeringmatters.reby

.media/2021/05/13/107-africa
-connecting-a-continent/.

36   **Four cables crossing the Congo
Canyon broke in August 2023:**
William Brederode, "Double Up the
Double Trouble: Undersea Cable
Break Was Worse Than Initially Let
On," News24, September 11, 2023,
https://subtelforum.com
/africa-cable-breaks-worse-than
-reported/.

36   **when four other cables broke:**
"Damage to Undersea Cables Is
Disrupting Internet Access Across
Africa," *The Economist*, March 21,
2024, https://www.economist
.com/middle-east-and-africa/2024
/03/21/damage-to-undersea-cables
-is-disrupting-internet-access
-across-africa.

36   **a cable can last as long as twenty-
five or thirty years:** "Submarine
Cable Frequently Asked Questions,"
Teleography.com, https://www2
.telegeography.com/submarine
-cable-faqs-frequently-asked
-questions.

39   **Divers still help guide a cable
from ship to shore:** "Landing Our
Grace Hopper Subsea Cable,"
YouTube video, uploaded
November 5, 2021, https://www
.youtube.com/watch?v=47e
_3WBCe-Q.

39   **that the city now hosts dozens
of data centers:** "Mumbai Data
Centers," Data Center Map, n.d.,

https://www.datacentermap.com
/india/mumbai/.

42   **Google had partnered with a
few other companies to lay a cable
across the Pacific:** Saul Hansell,
"Has Google Plans to Lay a
Pacific Cable?" *New York Times*,
September 21, 2007, https://
archive.nytimes.com/bits.blogs
.nytimes.com/2007/09/21/google
-plans-undersea-pacific-cable/.

43   **OTTs have funded more than
80 percent of new cable capacity:**
Melanie Mingas, "Subsea World
2020: How the OTTS Are
Changing Cable Finance," Capacity
Media, July 22, 2020, https://www
.capacitymedia.com/article
/29otcfa1zpmeiit2vnr40/news
/subsea-world-2020-how-the-otts
-are-changing-cable-finance.

44   **the four big tech giants
accounted for 69 percent of all
international capacity:** "The State of
the Network," TeleGeography, 2023,
https://www2.telegeography.com
/hubfs/LP Asscts/Ebooks/state-of
-the-network-2023.pdf.

45   **HMN Tech manufactured nearly
a fifth of all the subsea cables:** Joe
Brock, "U.S. and China Wage War
Beneath the Waves—Over Internet
Cables," Reuters, March 24, 2023,
https://www.reuters.com
/investigates/special-report/us
-china-tech-cables/.

45   **the US government blacklisted
Huawei:** Alexandra Alper, "U.S.

116   Actions Against China's Huawei," Reuters, n.d., https://www.reuters.com/graphics/USA-CHINA/HUAWEI-TIMELINE/zgvomxwlgvd/.

CHAPTER THREE

49   **US troops moved into position around many cable stations:** Nicole Starosielski, *The Undersea Network*, 120–125.

51   **including a repeater tower on the island of Kao:** "Case Study: Longest Microwave Link in a Mobile Phone Network," Aviat Networks, n.d., https://aviatnetworks.com/customer-stories/longest-microwave-link-mobile-network/.

53   **a sailor who'd berthed his yacht there temporarily:** "After Devastating Volcano Eruption, inReach® Devices Provide Critical Communication," Garmin, June 14, 2022, https://www.garmin.com/en-GB/blog/after-devastating-volcano-eruption-inreach-devices-provide-critical-communication/.

54   **has photos on his phone of the momentous day he took delivery:** DHL Tonga Agent's Post, April 10, 2022, https://www.facebook.com/permalink.php?story_fbid=pfbid02qAuYjL3XSNfWoCipx7qBWYA123LgoLUxBBtw3pTVGRTmrpVxxyzGwjeFXXDN85TXl&id=109137745097781.

55   **When another COVID lockdown was imposed in late March:** "Stricter Lockdown Rules to Be Enforced in Tonga," RNZ, March 18, 2022, https://web.archive.org/web/20220318144226/https://www.rnz.co.nz/international/pacific-news/463575/stricter-lockdown-rules-to-be-enforced-in-tonga.

55   **A Starlink costs around a quarter of a million dollars to manufacture:** LinkedIn post, Victor Tagborioh, https://www.linkedin.com/posts/victortag4_the-cost-of-manufacturing-a-single-starlink-activity-7160872445522903040-FTM_/.

55   **can be carried into orbit a couple of dozen at a time:** Stephen Clark, "SpaceX Pausing Launches of New-Generation Starlink Satellites," *Spaceflight Now*, March 23, 2023, https://spaceflightnow.com/2023/03/23/spacex-pausing-launches-of-new-generation-starlink-satellites/.

55   **Amazon has approvals to deploy 3,236 satellites:** FCC 20-102, Federal Communications Commission, July 30, 2020, https://docs.fcc.gov/public/attachments/FCC-20-102A1.pdf.

55   **Three Chinese projects—one of them named "Qianfan":** Magdalena Petrova, Jason Reginato, and Jeniece Pettitt, "The Chinese Competitors Aiming to Challenge SpaceX's Starlink Satellite System," CNBC, December 15, 2024, https://www.cnbc.com/video/2024/12/15

/chinese-competitors-aiming-to
-challenge-spacexs-starlink
-satellites.html.

60   **companies keep extra cable wound in depots located in ports around the world:** "Company History," SBSS, 2025.

60   **the Apia depot in Samoa, run by Subcom:** "TE SubCom to Build Cable Depot in Samoa," The Wire Association International, updated March 9, 2018, https://www .wirenet.org/news-categories/item /110-te-subcom-to-build-cable -depot-in-samoa.

60   **a 140-meter-long vessel named** *Reliance*: *Reliance*, Vessel Finder, accessed April 5, 2025, https:// www.vesselfinder.com/vessels/ details/9236494.

61   **the** *Reliance* **required seven drags:** Ulrich Spiedel, "Starlink for Tonga?" Bufferbloat, February 18, 2022.

66   **Nobody wants to be the** *Responder*: "Watch: Cable Laying Vessel Catches Fire and Sinks in East China Sea," *Marine Insight*, September 17, 2020, https://www .marineinsight.com/shipping -news/watch-cable-laying-vessel -catches-fire-and-sinks-in-east -china-sea/.

67   **there are just sixty-nine or so ships:** "Industry Report, 2023– 2024," Submarine Telecoms Forum, October 26, 2023, https://issuu.

com/subtelforum/docs/submarine _telecoms_industry_report _issue_12.

67   **had already filled its fleet's calendar for the next three years:** Dan Swinhoe, "The Cable Ship Capacity Crunch," DCD, December 6, 2022, https://www.datacenterdynamics .com/en/analysis/the-cable-ship -capacity-crunch/.

69   **committed $400 million to the company in 2023:** "KKR-backed Malaysian Group OMS Signs Contract for Cable-Laying Vessels with Dutch Firm," Reuters, October 27, 2024, https://www.reuters .com/business/media-telecom /kkr-backed-malaysian-group -oms-signs-contract-cable-laying -vessels-with-dutch-2024-10-28/.

70   **a marine maintenance coordinator named Surahtul Imran Ali:** Surahtul Imran Ali, LinkedIn profile, accessed April 5, 2025, https://www.linkedin.com /in/surahtulimran/.

73   **another earthquake wiped out the same domestic cable:** Kalafi Moala, "Parts of Tonga Without Internet After Cables Damaged and Starlink Ordered to Cease Operations," *The Guardian*, July 15, 2024, https://www.theguardian .com/world/article/2024/jul/16 /parts-of-tonga-without-internet -after-cables-damaged-and -starlink-ordered-to-cease -operations.

118   CHAPTER FOUR

75 "deliver seamless international connectivity to approximately 3 billion people": "2Africa Subsea Cable Makes First Landing in Genoa, Italy," Meta, April 14, 2022, https://about.fb.com/news/2022/04/2africa-subsea-cable-makes-first-landing-in-genoa-italy/.

75 Meta has seven other partners: "Partners," 2Africa, accessed April 5, 2025, https://www.2africacable.net/partners.

76 only 22 percent of people in sub-Saharan Africa were using mobile internet: https://www.worldbank.org/en/results/2024/01/18/digital-transformation-drives-development-in-afe-afw-africa.

76 a third of the population across Africa had any kind of broadband at all: "Digital Transformation Drives Development in Africa," World Bank Group, January 18, 2024, https://www.worldbank.org/en/results/2023/06/27/from-connectivity-to-services-digital-transformation-in-africa.

76 Meta claims Africa's economic activity will rise by $37 billion: Emeka Ekonkwo, "Meta's 2Africa Cable Landing a Windfall for African Economies," ItWeb Africa, March 22, 2024, https://itweb.africa/content/o1Jr5qxPebaqKdWL.

77 they control nearly half of oil and gas production in Africa: "Who Wins from Exploiting Africa's Oil and Gas?" *African Business*, August 1, 2022, https://african.business/2022/08/energy-resources/who-wins-from-exploiting-africas-oil-and-gas.

77 profits from the oil and gas tend to go overwhelmingly to these companies: "Who Profits from the New Oil?" Tagesspiegel, accessed April 5, 2025, https://interaktiv.tagesspiegel.de/lab/eacop-pipeline-through-east-africa-who-profits-from-the-new-oil/.

77 often as bribes to officials: Michael Race, "Glencore Ordered to Pay Millions Over Africa Oil Bribes," BBC, November 3, 2022, https://www.bbc.co.uk/news/business-63497376.

77 too meagerly as tax payments: "Africa Debate: Will Africa Ever Benefit from Its Natural Resources?" BBC, October 15, 2012, https://www.bbc.co.uk/news/world-africa-19926886.

77 two-thirds of the population lives on less than $2 a day: Ruth Olurounbi, "Nearly Two-Thirds of Nigerians Live on Less Than $2 a Day," *Bloomberg*, November 18, 2022, https://www.bloomberg.com/news/articles/2022-11-18/nearly-two-thirds-of-nigerians-live-on-less-than-2-a-day.

77 the blight of pollution in the Niger Delta: Macdonald Dzirutwe and Libby George, "Nigeria Oil Enters Unclear New Era After

Shell's Onshore Asset Sale," Reuters, January 29, 2024, https://www.reuters.com/markets/commodities/nigeria-oil-enters-unclear-new-era-after-shells-onshore-asset-sale-2024-01-29/.

77   **"System Honoring the Achievements of Rick Perry":** "Vodafone Connects UK to the World's Largest Subsea Cable System," Subtel Forum, June 7, 2024, https://subtelforum.com/vodafone-lands-2africa-cable-in-uk/. Also confirmed by Rick Perry. See the audio titled Rick Perry, timestamped 47m54s.

79   **ironically because of the very bandwidth issues that 2Africa was trying to address:** From the Bastian podcast.

80   **the cable landed in Nigeria:** "West Africa's MainOne Lands 2Africa Cable in Nigeria," ItWeb Africa, February 29, 2024, https://itweb.africa/content/Olx4zMka6oLv56km.

81   **thousands of fruit bats roost through the day:** David Esnault and Mariam Kone, "Emblems of a City, the Bats of Abidjan Face Troubled Future," Phys.org, March 17, 2021, https://phys.org/news/2021-03-emblems-city-abidjan-future.html.

81   **the *Ile de Sein*:** *Ile de Sein*, Marine Traffic, accessed April 5, 2025, https://www.marinetraffic.com/en/ais/details/ships/shipid

:171434/mmsi:226235000/imo:9247039/vessel:ILE_DE_SEIN.

82   **Ba has evangelized the kind of faster, more reliable access:** Ibrahima Ba, "Affordable Infrastructure Will Push Digitalisation" *The Nation*, November 22, 2021, https://thenationonlineng.net/affordable-infrastructure-will-push-digitalisation/#google_vignette.

82   **a landslide in the Trou sans Fond had ruptured all three cables:** "Rapport sur les ruptures de câbles sous-marins en Afrique de l'Ouest de 2024," Internet Society, April 18, 2024, https://www.internetsociety.org/fr/resources/doc/2024/rapport-sur-les-ruptures-de-cables-sous-marins-en-afrique-de-louest-de-2024/.

83   **Connectivity in Côte d'Ivoire plunged to 3 percent:** "Damage to Undersea Cables Is Disrupting Internet Access Across Africa," *The Economist*, March 21, 2024, https://www.economist.com/middle-east-and-africa/2024/03/21/damage-to-undersea-cables-is-disrupting-internet-access-across-africa.

83   **scores a measly 24 percent on the Internet Resilience Index:** "Internet Resilience," Internet Society, n.d., https://pulse.internetsociety.org/resilience.

83   **put the country's daily losses from a total outage at $24 million:** Netblocks Cost of Shutdown Tool, https://netblocks.org/cost/.

120   83   **full repairs took the better part of two months:** Niva Yadav, "WACS Subsea Cable Off West Africa Repaired After Landslide Damage," DCD, May 7, 2024, https://www.datacenterdynamics.com/en/news/wacs-subsea-cable-off-west-africa-repaired-after-landslide-damage/.

CHAPTER FIVE
87   **a short story by science-fiction writer Garry Kilworth:** David S. Garnett, ed., *Zenith: The Best in New British Science Fiction* (Sphere Books Ltd., 1989), https://archive.org/details/zenithbestin newbo000unse/mode/2up.

88   **Wiretapping was born in the same century as the telegraph cable:** Meyer Berger, "Tapping the Wires," *The New Yorker*, June 11, 1938, https://www.newyorker.com/magazine/1938/06/18/tapping-the-wires.

88   **was regularly on tapping duties:** "Cable Tappling Pod Laid by US Submarine Off Khamchatka," Interception Capabilities 2000, April 1999, https://irp.fas.org/eprint/ic2000/ic2000.htm#_Toc448565536.

89   **devices known as "intercept probes" can be installed:** "British Spy Agency Taps Cables, Shares with U.S. NSA -Guardian," Reuters, June 21, 2013, https://www.reuters.com/article/technology/british-spy-agency-taps-cables-shares-with-us-nsa-guardian-idUSL5N0EX3JA/.

89   **the US National Security Agency (NSA) and the UK Government Communication Headquarters (GCHQ):** Ewan MacAskill, Julian Borger, Nick Hopkins, Nick Davies, and James Ball, "GCHQ Taps Fibre-Optic Cables for Secret Access to World's Communications," *The Guardian*, June 21, 2013, https://www.theguardian.com/uk/2013/jun/21/gchq-cables-secret-world-communications-nsa.

89   **used intercept probes at least as early as 2002:** "Whistle-Blower's Evidence, Uncut," *Wired*, May 22, 2006, https://www.wired.com/2006/05/whistle-blowers-evidence-uncut/.

89   **described the NSA's installation of this equipment:** Justin Elliott, "Does the NSA Tap That? What We Still Don't Know About the Agency's Internet Surveillance," *Pro Publica*, July 22, 2013, https://www.propublica.org/article/what-we-still-dont-know-about-the-nsa-secret-internet-tapping.

89   **the US has replaced the *Parche* with a submarine named after Jimmy Carter:** "New Nuclear Sub Is Said to Have Special Eavesdropping Ability," *The New York Times*, February 20, 2005, https://www.nytimes.com/2005/02/20/politics/new-nuclear-sub-is-said-to-have-special-eavesdropping-ability.html.

89   **interviewed an expert who offered another technique:** Olga Khazan, "The Creepy, Long-Standing Practice of Undersea Cable Tapping," *The Atlantic*, July 16, 2013, https://www.theatlantic.com/international/archive/2013/07/the-creepy-long-standing-practice-of-undersea-cable-tapping/277855/.

90   **"We kill people based on metadata":** Bruce Schneier, "NSA Doesn't Need to Spy on Your Calls to Learn Your Secrets," *Wired*, March 25, 2015, https://www.wired.com/2015/03/data-and-goliath-nsa-metadata-spying-your-secrets/.

91   **moving to restrict the American activities of Huawei:** Charles Arthur, "China's Huawei and ZTE Pose National Security Threat, Says US Committee," *The Guardian*, October 8, 2012, https://www.theguardian.com/technology/2012/oct/08/china-huawei-zte-security-threat.

91   **Primarily, this referred to cyber-espionage:** Noah Berman, Lindsay Maizland, and Andrew Chatzky, "Is China's Huawei a Threat to U.S. National Security?" Council on Foreign Relations, updated February 8, 2023, https://www.cfr.org/backgrounder/chinas-huawei-threat-us-national-security.

91   **via an executive order:** "Executive Order on Establishing the Committee for the Assessment of Foreign Participation in the United States Telecommunications Services Sector," executive order, April 4, 2020, https://trumpwhitehouse.archives.gov/presidential-actions/executive-order-establishing-committee-assessment-foreign-participation-united-states-telecommunications-services-sector/.

91   **influences the granting of licenses to land submarine cables on US soil:** "Q&A: Team Telecom Review and National Security," *Financier Worldwide*, September 2023, https://www.financierworldwide.com/qa-team-telecom-review-and-national-security.

92   **was forced to drop China's HMN Tech as its contractor:** Nicole Starosielski, *The Undersea Network*, 61.

92   **approvals could take as long as six hundred days:** Nicole Starosielski, *The Undersea Network*, 55.

92   **recommended that a cable connecting Cuba to the US be nixed:** "Team Telecom Recommends the FCC Deny Application to Directly Connect the United States to Cuba Through Subsea Cable," U.S. Department of Justice, November 30, 2022, https://www.justice.gov/opa/pr/team-telecom-recommends-fcc-deny-application-directly-connect-united-states-cuba-through.

122

93 **denied permission to lay a cable between Hong Kong and the US:** "Team Telecom Recommends That the FCC Deny Pacific Light Cable Network System's Hong Kong Undersea Cable Connection to the United States," US Department of Justice, June 17, 2020.

93 **US forced the contract to be flipped to SubCom:** Joe Brock, "U.S. and China Wage War Beneath the Waves—Over Internet Cables."

93 **laid by Meta and Chinese telecom companies:** "HKA," Submarine Cable Networks, n.d., https://www .submarinenetworks.com/en /systems/trans-pacific/hka.

94 **one from Japan, one from Korea, and SBSS:** "Ship/ROV Operators," Yokohama Zone, accessed April 5, 2025, https://yokohamazone.com /ship_ROV%20Operators.html.

94 **noticed how carefully OMS describes itself on its website:** OMS Group homepage, https:// opticmarine.com/about-2/.

94 **delaying permits to any cables planned through the South China Sea:** Tsubasa Suruga, "Asia's Internet Cable Projects Delayed by South China Sea Tensions," *Nikkei Asia*, May 19, 2023, https://asia.nikkei .com/Business/Business-Spotlight /Asia-s-internet-cable-projects -delayed-by-South-China-Sea -tensions.

94 **Some cables have had to be rerouted entirely:** Rebecca Tan, "Escalating Contest over South China Sea Disrupts International Cable System," *Washington Post*, October 3, 2024, https://www .washingtonpost.com/world/2024 /10/03/south-china-sea -underwater-cables/.

93 **US lays its own military cables:** Joe Brock, "SPECIAL REPORT— Inside the Subsea Cable Firm Secretly Helping America Take On China," Reuters, July 6, 2023, https://www.reuters.com/article /markets/commodities/special -report-inside-the-subsea-cable -firm-secretly-helping-america -take-on-chi-idUSL1N38P1Z3/.

95 **all these permitting complications have increased the expense and duration of projects:** Sunil Tagare, "The US Government Is Wrong About Submarine Cables," LinkedIn Pulse, November 12, 2024, https://www.linkedin.com/pulse /us-government-wrong-submarine -cables-sunil-tagare-qxyrc/.

96 **another army facility, now abandoned, called the Iron Fort:** "Iron Fort," Taiwan Waves of Wonder, https://eng.taiwan.net.tw /m1.aspx?sNo=0002127&id=6517.

97 **On paper, the ships were a fishing boat and a cargo vessel:** Jono Thomson, "Matsu Undersea Cable Repaired Ending 50 Day Internet Outage," *Taiwan News*, March 31,

2023, https://www.taiwannews
.com.tw/news/4852575.

97   **the Chinese navy so frequently:**
Diego Laje, "Uncovering China's
Fishy Activities at Sea," *Signal*,
March 1, 2023, https://www.afcea
.org/signal-media/intelligence
/uncovering-chinas-fishy
-activities-sea.

97   **had been damaged at least
twenty times:** Jono Thomson,
"Matsu Undersea Cable Repaired
Ending 50 Day Internet Outage."

97   **its eastern flank is seismically
unstable:** W. S. Chen, Y. M. Wu,
P. Y. Yeh, Y. X. Lai, S. S. Ke, M. C.
Ke, and C. Y. Yang, "Insights Into
the Seismogenic Structures of
the Arc-Continent Convergent
Plate Boundary in Eastern
Taiwan," *Terrestrial, Atmospheric,
and Oceanic Sciences* 35, no. 13
(2024), https://doi.org/10.1007
/s44195-024-00065-7.

100   **who was serving as Taiwan's
deputy minister of digital affairs:**
"Deputy Minister Herming
Chiueh," MODA Ministry of Digital
Affairs, August 27, 2022, https://
moda.gov.tw/en/aboutus/principal
-officers/deputy-minister-chiueh
/993.

101   **Seven hundred ground-based
satellite receivers:** Fern Hinrix,
"Building Resilience in Taiwan's
Internet Infrastructure from
Geopolitical Threats," note 16.

101   **heavy criminal penalties for
anyone who damaged subsea cables:**
Shelley Shan, "Undersea Cable
Damage Penalties May Be Raised,"
Taipei Times, March 30, 2023,
https://www.taipeitimes.com
/News/taiwan/archives/2023/03
/30/2003797002.

102   **no effective, coherent
body of law to hold responsible:**
Amy Paik and Jennifer Counter,
"International Law Doesn't
Adequately Protect Undersea
Cables. That Must Change,"
Atlantic Council, January 25, 2024,
https://www.atlanticcouncil.org
/content-series/hybrid-warfare
-project/international-law-doesnt
-adequately-protect-undersea
-cables-that-must-change/;
Also: James Kraska, "The Law
of Maritime Neutrality and
Submarine Cables," *EJIL: Talk!*,
July 29, 2020, https://www.ejiltalk
.org/the-law-of-maritime
-neutrality-and-submarine
-cables/.

104   **already rumors were fizzing:**
Ingrid Lunden, "Meta Plans to Build
a $10B Subsea Cable Spanning the
World, Sources Say," *TechCrunch*,
November 29, 2024, https://
techcrunch.com/2024/11/29/meta
-plans-to-build-a-10b-subsea
-cable-spanning-the-world
-sources-say/; Jess Weatherbed,
"Meta Is Building the 'Mother
of All' Subsea Cable," *The Verge*,
November 29, 2024, https://www
.theverge.com/2024/11/29

124   /24308746/meta-10-billion
-global-subsea-cable-project.

104   "one hell of a cable": Sunil
Tagare, "Map of Meta's 'W' Cable,"

LinkedIn Pulse, October 22, 2024,
https://www.linkedin.com/pulse
/map-metas-w-cable-sunil-tagare
-blanc/.

Columbia Global Reports is a nonprofit publishing imprint from Columbia University that commissions authors to produce works of original thinking and on-site reporting from all over the world, on a wide range of topics. Our books are short—novella-length, and readable in a few hours—but ambitious. They offer new ways of looking at and understanding the major issues of our time. Most readers are curious and busy. Our books are for them.

If this book changed the way you look at the world, and if you would like to support our mission, consider making a gift to Columbia Global Reports to help us share new ideas and stories.

Visit globalreports.columbia.edu to support our upcoming books, subscribe to our newsletter, and learn more about Columbia Global Reports. Thank you for being part of our community of readers and supporters.